Good Farmers

GOOD FARMERS

Traditional Agricultural Resource Management in Mexico and Central America

Gene C. Wilken

University of California Press
Berkeley Los Angeles London

University of California Press
Berkeley and Los Angeles, California

University of California Press, Ltd.
Oxford, England

First Paperback Printing 1990

Library of Congress Cataloging-in-Publication Data

Wilken, Gene C.
Good farmers.

Includes index.
1. Agricultural resources—Mexico—Management.
2. Agricultural resources—Central America—Management.
3. Agriculture—Mexico. 4. Agriculture—Central
America. I. Title.
S451.7.W55 1987 338.1′0972 86-16146
ISBN 0-520-07205-7

Printed in the United States of America

1 2 3 4 5 6 7 8 9

The paper used in this publication meets the minimum requirements of American National Standard for Information Sciences—Permanence of Paper for Printed Library Materials, ANSI 239.48-1984. ♾

Dedicado a los Señores
Don Bernardino Nophal Corona
Don Pablo Ahguech Rodríguez
Don Pedro Gonón
Don Apolinar Rodríguez
Don Juan Vásquez Santiago
Campesinos, Maestros, Amigos

y a todos los otros quienes cuidan los escasos recursos que les proporcionan el ambiente y la sociedad, y quienes con paciencia y comprensión me han revelado la utilidad y la belleza de su trabajo.

Contents

Preface

My reasons for writing this book about traditional agricultural technology in Mexico and parts of Central America are several. The first reason is to document the methods used by some farmers to manage agricultural resources under severe constraints of energy, technology, and scale. In this regard the book offers a useful catalog of practices employed in one part of the nonindustrial world.

The second reason is to formulate some principles of traditional resource management so that agricultural systems from different areas of the world can be compared. The constraints found in Middle America are common to traditional agriculture everywhere, and the solutions, even in widely separated regions, often share basic features. Although practices differ, their goal is to achieve similar results, and their effectiveness can be evaluated within particular physical, biological, and social contexts. Thus, rather than assuming the superiority of one set of agricultural methods over another, this book identifies the challenges faced by Middle American farmers, then examines the solutions they have developed.

A third reason for writing this book is to argue that the challenges of managing agricultural resources—soil, water, slope, climate, and space—to enhance crop plant environments are common to modern as well as to traditional farmers everywhere and require comparable managerial responses. Most of the factors that characterize traditional agriculture are technological: small size of operation and reliance on human or animal power and local information. To Western readers, some of the tools and techniques may seem quaint, even primitive. But as good scientists we eschew such ethnocentric evaluations and ask instead: "Do these methods work?" "Can we learn from them?" "Could they be used elsewhere?" In a broad sense, the solutions as well as the challenges are common, and in the end we deal not with traditional or modern or even appropriate technology, but simply with technology by which, with more or less success, farmers manage field environments for the benefit of their crops.

The book is organized around five major resource categories. After outlining the scope of the study in Chapter 1 and discussing the sources and uses of energy in traditional agriculture in Chapter 2, I examine the management of soil (Chapters 3 to 5), slope (Chapters 6 and 7), water (Chapters 8 to 10), climate (Chapter 11), and space (Chapter 12). In Chapter 13, I evaluate the prospects for traditional resource management in the future.

This method of organization raises a major problem that was never completely resolved. Because many traditional techniques are multipurpose, organizing by resource risks repetitive descriptions of the same practices. For example, terraces basically control field gradients and surface flows on sloping fields. But they also increase infiltration and are necessary for irrigation and soil management. Thus, in a resource-by-resource treatment, terraces must frequently reappear, and with them the question of how much detail to review. The alternative of organizing by technique, however, would risk the even greater problem of reviewing resource characteristics with each practice. I adopted the resource approach as the lesser of evils. To minimize repetition, I discuss details of particular techniques on first mention and review them only briefly thereafter.

More people helped write this book than can reasonably be listed here. A few have been especially helpful. I am happy to acknowledge my debts to Woodrow Borah, Charles E. Bussing, William M. Denevan, Steven R. Gliessman, Efriam Hernández X., David H. Miller, James J. Parsons, Donald E. Vemeer, and Paul J. Zinke. Each not only practices his respective discipline but also has helped shape it and, by so doing, has influenced the work and thought of countless others. Field work was made possible by financial assistance from the Foreign Area Fellowship Program of the National Academy of Sciences/American Council of Learned Societies, the National Science Foundation, and the Wenner-Gren Foundation for Anthropological Research. My wife, Karen, and children, Oliver, Kirsten, and Harold, shared the joys and sorrows of field work and writing.

Gene C. Wilken
Fort Collins, Colorado
29 November 1986

1

Studying Traditional Resource Management

Definitions

Resource Management

Studies of traditional agriculture and resource management are still fairly rare. Perhaps these practices are so familiar that they have escaped the systematic investigations that normally accompany a body of technology. Whatever the reason, it is necessary to start fresh, to define the phenomena we shall be discussing, and to establish the scope to this study. To do this we must also examine farming in relation to the environment since the practices are best understood as they influence the natural processes that occur in cultivated fields.

All farming systems attempt to direct or increase flows of energy and materials into the production of useful products.[1] But farming is not manufacture: the plants themselves combine elements from the environment into new forms without the direct application of labor or other human inputs. Regardless of its methods, farming depends on

[1]M.J.T. Norman (*Annual Cropping Systems in the Tropics*, pp. 2–10) notes the many confusing uses of the phrase *farming system*. I accept his succinct definition of a farming system as the "pattern of resources and processes of resource use in a farming unit," in which resources include natural, human, and capital resources and incipient products, and the farming unit in any specific instance can be a geographical or economic unit or a unit displaying a particular pattern of resource use.

natural flows of energy and materials. Farmers can change these flows only indirectly, by modifying the plant environment.

It has become popular to define the human role in this process as ecosystem modification. This useful approach avoids static definitions based on crop type or tools and instead stresses crop-environment relationships. But ecosystem modification in agriculture is not random; it is directed toward the production of goods that have value, whether consumed by the farmer or sold. Various parts of a farming system, including elements of the physical environment, have utility because of their potential contribution to the productive process. The degree to which these potentials are realized is a measure of technological competence. The notions of production and value create a basis for system comparison and evaluation.

Concepts such as value and utility are foreign to strict ecological analyses in which system success is measured in terms of stability, adaptability, or survival. But for farming systems it is appropriate to recognize these cultural and economic aspects by substituting the concept of resource management for ecosystem modification in order to convey the sense of purposeful production of items of value.[2]

There are three elements involved in the process of resource management: the crop plants; the environment, including solar energy in its various transformations; and the farmers who, with their knowledge, tools, and institutions, actually manage the modification and maintenance of the ecosystem.

Although draft and barnyard animals are integral to most Middle American small-farm operations, they are customarily fitted into human or plant environments, such as house yards or fallow fields, rather than environments specially modified for their use. Because their needs influence to only a minor extent the resource management practices discussed here, animals will figure in this study primarily as power sources for field operations.

The great achievements of traditional plant management in selection, domestication, and genetic manipulation also will not be reviewed here.[3] With few exceptions, plant characteristics change so slowly that they cannot be controlled by individual farmers. However, opportunities for significant modification of the environment are

[2]For a discussion of the managerial approach, see Harold C. Brookfield, "New Directions in the Study of Agricultural Systems in Tropical Areas." This approach is distinct from contemporary agroecosystem research, which stresses ecological relationships (for example, J. L. Harper, "Agricultural Ecosystems"; Orie L. Loucks, "Emergence of Research on Agro-Ecosystems").

[3]Peter J. Ucko and G. W. Dimbleby, eds., *The Domestication and Exploitation of Plants and Animals.*

available to all farmers, regardless of time, place, or level of technology. In fact, farming systems are essentially collections of techniques for modifying plant environments.[4]

Similarly, the physical environment can be considered relatively constant. Although changes in climate, soil, and slope are important and account for different crop mixes and management practices, these elements remain quite fixed in a given area. Traditional techniques to manipulate the environment are closely attuned to known conditions and take into account expected variations. Even extraordinary events (hazards) are in a real sense part of the familiar environment, to be endured or meliorated according to the availability of defensive strategies.[5]

The specific focus here is on the environmental elements that restrain plant development or agricultural activities, especially those that can be managed by available techniques. We shall examine five major agricultural resources: soil, water, surface geometry, climate, and space. Although water originates as precipitation, it is considered separately from other climatic elements because of its distinctive nature once it is on or below the surface.[6]

Divisions among these resources are seldom clear cut. Most resources interact such that it is almost impossible to isolate them or their management techniques, and many practices are multifunctional. In the end, assigning a practice to a particular resource was based on my own or the farmers' opinions of what seemed to be the main purpose.

Traditional Farming

These resources and, indeed, most of the major crop plants are not exclusive to any particular type of agricultural operation. What then distinguishes a farming system in Kansas or the Ukraine from one in Guatemala or Mexico? The products, for example grains of wheat or maize, are for all practical purposes indistinguishable, as are plant responses to environmental factors such as soil moisture or nitrogen levels. Clearly, the distinctions between farming systems must lie in differences of resource management techniques and of scale.

[4]Even the high-yielding plant varieties of the Green Revolution could better be called high-response varieties because their distinguishing characteristic is that they respond to drastic modifications in their environment, especially the use of chemical fertilizers. N. W. Pirie, "Using Plants Optimally."

[5]Gene C. Wilken, *Agroclimatic Hazard Perception, Prediction, and Risk-Avoidance Strategies in Lesotho.*

[6]David H. Miller, *Water at the Surface of the Earth.*

Techniques

Resource management is a concept developed in industrially advanced societies, where scientific understanding of physical and chemical processes joined with modern concepts and equipment to produce methods for maximizing production or minimizing the deleterious effects of resource exploitation. But resource management, or modification of the plant environment, is an integral part of any farming system regardless of its complexity or technological stage. If we wish to distinguish between systems, we must look for differences in the techniques of resource management, specifically in the means by which information is transferred, and in the tools and energy sources by which the environment is modified.

Generation and dissemination of information tend to be highly formalized and visible in modern agriculture. Laboratories and extension services, factories and sales programs transfer information from many sources to individual farms through a variety of media.[7] In contrast, traditional systems depend primarily on individual or local group knowledge of resource characteristics and management methods. In fact, the transfer of knowledge based on experience from one generation to another is the essence of the term *traditional*.[8] Information tends to be transmitted verbally and informally, through individual exchanges, village meetings, and local markets (where one meets merchants and farmers from other areas). Although such local systems suffer from restricted sources of information, they do make all options available to farmers within a given area.

Traditional information systems include two broad aspects of resources: their characteristics and their management.[9] Knowledge of resource characteristics is often formalized in classification systems. Dividing the biophysical elements into groups with similar or associated traits allows farmers to deal in a meaningful way with the otherwise unwieldy forces of nature. Such traditional agricultural resource classification tends to be utilitarian, emphasizing qualities related to suitability for certain crops or uses or to overall productivity.

[7]D. McArthur ("Information Sources for Farmers from Outside Their Farms") speaks of the "massive quantity and range of information" available to Canadian farmers from the mass media, commercial sources, and agricultural agencies, as well as from neighbors and friends.

[8]On this subject Theodore W. Schultz (*Transforming Traditional Agriculture*, pp. 33, 37) says: "The concept of traditional agriculture implies long-established routines with respect to all productive activities. . . . The factors of production on which they [traditional farmers] depend are known through long experience and are in this sense 'traditional.' "

[9]A third major aspect, resource ownership or control, falls outside the scope of this study.

Modifiers such as *more* and *less* permit farmers who share an understanding of basic resource qualities to express fine gradations. But in addition to inherent characteristics, classification systems describe propensities for change or manipulation, which in turn reflect levels of technology and management skills. For example, not only are the inherent chemical, textural, and fertility qualities of a soil important, but so too is its amenability to plowing or irrigation.

When we think of the kinds of resources that traditional farmers describe in this way, examples of climate and vegetation classification come first to mind, perhaps because these elements are apparent to everyone and have most attracted the attention of outside observers. But traditional farmers also classify soils, water quality, silts, manures—in fact, almost any variable resource that affects management or crop performance.[10]

Differences in tools and energy sources offer an even more obvious basis than differences in ways information is transferred for distinguishing resource-management systems. Certainly the impressive power units and specialized equipment of industrialized farmers stand in striking contrast to the simple hoes and plows of their traditional counterparts. But, as I have noted, the basic requirements of resource management, and thus the functions of the tools, remain essentially the same. Energy sources and applications are a better distinguishing characteristic than tools and will be explored in the next chapter.

Scale

Mention of scale in agriculture initially evokes comparisons of farm size, from the tiny fields of the less-developed world's small farms[11] to the enormous holdings found in technologically advanced sectors and societies. But small or large, farm refers to the surface area of holding and is after all only an index of what actually is a three-dimensional space in which energy, water, and minerals combine into products of value. Because some of the most distinctive management methods of traditional agriculture are a result of limited space, it seems reasonable

[10]Brent Berlin, Dennis E. Breedlove, and Peter H. Raven, "General Principles of Classification and Nomenclature in Folk Biology." Despite the complaint by these same authors elsewhere that reports of ethnobiological knowledge are sketchy ("Preface" to *Principles of Tzeltal Plant Classification*), folk taxonomies of plants and animals are relatively common. For example, compare the number of entries under the various plant and animal categories in Harold C. Conklin, *Folk Classification*.

[11]More than 80 percent of all holdings reported in the last world census were smaller than 5 ha. Food and Agriculture Organization, *1970 World Census of Agriculture*, pp. 22ff.

to consider space a scarce resource and to examine methods for its optimum use.

In addition to size of holding, traditional farms are often uniquely local; the region where they exist supplies most farm resources and absorbs most farm products. Associated social and economic institutions, like the farming systems themselves, tend to be local,[12] in striking contrast to the national and even international links of large-scale farming operations. Although weakened by alternative uses, the phrase *small scale* is useful for identifying these spatially and economically isolated farming operations.

Qualifications

The regimens of farming penetrate to the core of those societies that practice it, profoundly influencing patterns of spatial, environmental, and social interaction. With reason we speak of agrarian societies, noting in the name the unity of work and life. The interlocking social, economic, and ecological webs have produced many specific and general adjectives for farming, some of which confuse more than clarify. For example, these adjectives can emphasize the spatial aspects of farming (shifting, extensive), tools (hoe, plow), specific environments (flood plain, hillside), sociopolitical relationships (peasant), economics (subsistence, commercial), and processes for transmitting skills and knowledge (traditional). Because global technological differences dominate our thinking, it is also popular to pose dichotomies such as nonindustrial/industrial or even paleotechnic/neotechnic.[13] All these terms (except such ethnocentric and judgmental contrasts as primitive/modern) can be useful as long as they match the emphasis of a particular study.[14] But because we shall deal here with farming systems that depend primarily on individual or group knowledge of resource characteristics and on human or animal power for resource management, it is appropriate to describe them as traditional, nonindustrial, or

[12]Although Robert Redfield's self-sufficient *Little Community* would seem to offer a convenient model, its shifting cultivation economy lacks the integrating regional transfers of resources and farm products. Scale differences are explored in Gene C. Wilken, "The Fallacy of Labour/Land Comparisons of Agricultural Productivity."

[13]R. S. Loomis ("Traditional Agriculture in America") reviews some of these terms. *Paleotechnic* and *neotechnic* were first proposed by Patrick Geddes in 1915 *(Cities in Evolution)*. Subsequent writers added scale and energy concepts and applied the terms to agriculture.

[14]Brookfield ("New Directions") suggests basing primary descriptions of agricultural systems on agronomic considerations of farming techniques and management. For a recent review of classification methods see Jerzy Kostrowicki, "Agricultural Typology Concept and Method."

in some cases labor intensive in addition to the already-established term *small scale.*

Yet even with these operational definitions we must accept the continuity of farming and recognize that most traditional systems include some modern elements, just as most modern systems retain vestiges of the traditional,[15] and that between the two polar models lie all possible combinations. For all our care in establishing definitions, the proposed terms are more abstractions than descriptions, and individual examples are simply more, or less, traditional.

Scope and Limitations

Having defined traditional resource management and identified those characteristics that will concern us here, we can proceed to an examination of particular techniques. But first we must specify the goals and limits of the study.

This book reviews traditional agricultural resource management in Middle America. Examples are from my own field work, mostly in Mexico and Guatemala, and emphasize sophisticated, labor-intensive methods that demonstrate the potential of traditional technology. The sample admittedly is limited, and I have tried to commensurately limit generalization. Such restraint was not easy. Distressingly few methods, analyses, or theories exist in this scientific, technical, and scholarly field, despite its size and promise. In such a vacuum it is easy to erect conjectural edifices that exceed the data.

These data are both descriptive and numerical. Those that pertain to environmental elements and relationships are widely applicable; those that deal with cultural practices are less so. But, even here, the common characteristics of scale and energy link traditional farming systems the world over and impart a degree of universality to the findings.

I have not attempted to calculate rates of return from my field data on time, motion, weights, volumes, and labor (which are supplemented by figures from the literature and other sources). Crop yields are notoriously elusive in traditional agriculture, and the additional obstacle of assigning yield increases to particular management practices proved insurmountable. Presumably the effort (investment) of rebuild-

[15]William M. Denevan, "Latin America." H. J. Hopfen (*Farm Implements for Arid and Tropical Regions,* p. 4) observes that "even today on motorized holdings, man still spends two-thirds to three-quarters of his working time with hand implements."

ing a terrace or cleaning a drain eventually is repaid in reduced operating (labor) costs or higher yields. But data of sufficient fineness to permit such evaluations were not available.[16]

In addition, the returns from resource management depend as much on environmental conditions or limitations as on the management system itself. The greatest results come from relieving the most restrictive situations. For example, without irrigation, arid zones and seasons may be completely barren. Similarly, drainage and terracing can reclaim otherwise uncultivatable boggy soils and precipitous slopes. The potentials for water or slope management are greater in these extreme situations than they are for lands only marginally dry, saturated, or steep, whereas the labor investments and operating costs may not be proportionately larger.

Finally, the returns from traditional agriculture cannot be measured only by productivity, the preferred standard in market-dominated farming systems. In addition to quantity, traditional farmers may feel that variety, quality (also locally defined), risk reduction, and even prestige justify labor expenditures even though such investments may not have tangible returns.

All the resource-management practices reported here are taken from contemporary Middle America (Map 1-1) with supporting observations and references from other times and places. That many of these practices are of known or suspected pre-European origin is incidental to this study. It is also not surprising that most of these farming systems support dense rural populations; the association between population and resource management has been noted for many areas of the world.[17] Dense rural populations supply both the need for high productivity and the labor with which to achieve it.

This catalog of Middle American resource-management practices is by no means complete. I did not survey much of the region, and, even in the limited study areas, I suspect I missed more than I found. The reasons lie in both the practices and the observer. Traditional technology is closely attuned to local environmental and social conditions and varies, sometimes dramatically, from one area to another. A single small valley may shelter management techniques unknown elsewhere. In

[16]Long- and short-term benefit/cost considerations also cloud the issue. Traditional farmers must realistically estimate how well future benefits justify present labor expenditures. Although industrialized farmers face similar decisions, they often are assisted by subsidies or tax credits.

[17]For example, William Allen, *The African Husbandman;* Ester Boserup, *The Conditions of Agricultural Growth;* Clifford Geertz, *Agricultural Involution;* F. H. King, *Farmers of Forty Centuries;* Gary A. Klee, ed., *World Systems of Traditional Resource Management,* especially the chapters on East and South Asia.

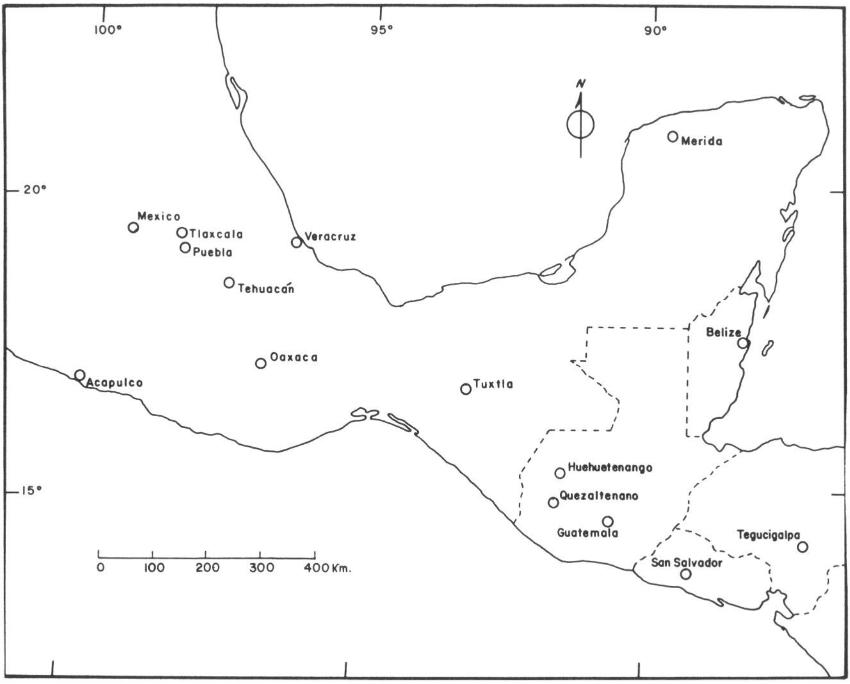

Map 1-1. General location of major areas mentioned in this study.

addition, traditional management forms are small and unobtrusive, seldom evidenced by specialized tools or impressive structures. The outside observer, accustomed to agroecosystem manipulation on a grand scale, can easily miss the significance of a row of stones on a hillside or a shading branch over a seed bed.

But these are the tools and structures with which traditional farmers control flows of water and energy in their small plots. Do they work? Of course they do! A long record of successful farming, often under difficult conditions, attests to their effectiveness. Are they efficient? As we have seen, that is a more difficult question; the answer depends on one's objectives. Is production paramount? Income? Employment? Ecological stability? Security? If the last two are indices of overall success, then traditional systems should receive high marks. Some of the practices to be discussed have been in use for centuries, even millennia, and have been the means by which their farmer/practitioners survived and even prospered. The fact that they still are in use is strong evidence for a social and ecological stability that modern, mechanized systems could well envy.

2

Energy Needs, Sources, and Strategies

Energy Needs

Agriculture depends on two major sources of energy: direct solar energy that drives the natural processes of photosynthesis and evaporation, and fossil fuel and human and animal food energy that provides the power for plant and resource management. Since the objective of management is to channel solar energy through the plants into desired products, farmers essentially spend energy from one source and collect it from another. This relationship is often shown as an input-output ratio of energy expended to energy harvested. Agricultural efficiency can be increased by either increasing harvested energy or reducing management energy.

These energy relationships apply to all types of farming. What distinguishes traditional from modern or industrial farming are the sources and quantities of energy used and the types of tools that convert power to useful work. In this chapter some of the energy sources and requirements of traditional management as well as the strategies farmers use to accomplish necessary tasks with their meager energy supplies are considered. Additionally, some of the most important characteristics and constraints of traditional agriculture are reviewed.

Like all living things, humans are energy consumers, requiring about one million kilocalories per person each year. These calories are derived almost entirely from plants or from animals that eat plants and thus represent solar energy transformed by photosynthesis into organic forms.

Although the amount of solar energy that enters the earth's many ecosystems is stupendous in the aggregate, most of it is reflected by clouds or the earth's surface, or goes to heat the environment and evaporate water. Less than one percent is photosynthesized to produce plant biomass, and an even smaller amount winds up in the edible parts of plants.

Because the composition and density of the plant cover where this transformation takes place vary in response to environmental factors, principally temperature and moisture, the rate at which solar energy is transformed and stored in organic form differs widely from place to place. The procurement of these unevenly scattered, edible calories is a major challenge for all large animals.

Of three basic responses to this challenge, two are available to all animals, the third only to humans. The most widespread strategy is foraging—ranging widely in daily grazing or hunting patterns and seasonal migrations. The human equivalent of foraging is hunting/gathering, in which bands of individuals gather plants and animals locally on a daily or weekly basis and visit the several ecosystems of their territories at different seasons of the year.[1] A variation is nomadic herding, in which grazing or browsing animals do the actual foraging and in turn are harvested by the herders.

The second strategy is for consumers to stay in one place while some mechanism, usually moving water, brings the calories to them. Many amphibians employ this strategy, waiting in concealment or camouflage until an incautious prey passes by. A human parallel is to adopt an advantageous riverbank or seaside location and harvest stream- or tide-borne foods. Because most plants are anchored, the catch is usually limited to animals, and carbohydrates still must be collected by foraging.[2]

Although both these strategies may produce significant modification of the environment, they can hardly be called resource management in the sense of purposeful control of energy, moisture, and nutrient flows in ecosystems. Even deliberate efforts such as seasonal prohibition of

[1]Bruce Winterhalder and Eric Alden Smith, *Hunter-Gatherer Foraging Strategies.*

[2]This distinction between foraging and sedentary hunters is perhaps too sharply drawn. Many groups employ a combination of techniques: settled fisherfolk also pursue game, and wide-ranging hunters also set traps. For an account of active and passive near-shore hunting and back-country gathering, see Bernard Nietschmann, *Between Land and Water.*

hunting game or gathering only in specified areas are more passive conservation measures than active resource management.[3]

The third strategy is uniquely human: modifying the environment to increase flows of energy, moisture, and nutrients into valued plants. This process is called farming and consists of several elements: substitution of economic for noneconomic plants; selection of plants for desirable characteristics, especially abnormal development of edible parts; and modification of the physical environment to favor the growth and development of desirable plants (including intervention against pests, disease, and weeds).

The three strategies are by no means mutually exclusive: clever hunters/gatherers/herders/fishers/farmers can and may employ several or all of them in a single economic system. Neither are there clear advantages to any one approach.[4] Although the systems that do not involve resource management often enjoy favorable energy input/output ratios, they are efficient only under certain environmental and demographic conditions.[5] Furthermore, simple biological maintenance has never been the sole objective of human societies, and energy efficiency is but one measure of the success of an economic system.

Compared with hunting and gathering or herding, most farming systems are less efficient; farming produces more food per unit of area, but only with higher costs and consequently lower returns per unit of time.[6] Farm work is notoriously arduous: workdays are long and the tasks are among the most strenuous and repetitive in human experience.[7] But just where is human and animal energy expended in farming?[8] Most of it goes to control various elements of the crop plant environment. Although the flows of moisture and nutrients within ecosystems, including agroecosystems, are driven primarily by solar energy, farmers attempt to control these flows, and thus plant growth and development,

[3]Gary A. Klee ("Oceania") reviews marine resource conservancy in the South Pacific.

[4]Bryony Orme ("The Advantages of Agriculture") reviews possible reasons for the transition to farming after millennia of successful hunting/gathering. Among contemporary traditional farmers in Middle America residual hunting and gathering are common, especially of plants and animals that thrive in human-modified ecosystems. Gene C. Wilken, "The Ecology of Gathering in a Mexican Farming Region."

[5]Gerald Leach, *Energy and Food Production,* p. 7.

[6]Colin Clark and Margaret Haswell, *The Economics of Subsistence Agriculture,* pp. 103–11; Leach, *Energy and Food Production.* E. C. Pasour, Jr. ("Economic Efficiency") examines the subjective nature of efficiency.

[7]Over the centuries the popular image of farm work has alternated from romantic husbandry to plodding drudgery.

[8]Inputs of capital or labor often are discussed without reference to methods of application, as if either could be introduced directly into what are essentially biophysical systems.

Table 2-1. Energy Expenditures in Agricultural Resource Management

Activity	Averages (and Ranges)[a] (kcal/min)	
Clearing		
Tree felling	8.2	(—)
Clearing shrubs and grass	7.1	(5.8–8.4)
Bush clearing	6.1	(—)
Preparing fields		
Plowing		
With horse	—	(5.1–5.9)
With tractor	4.2	(—)
Ridging	9.5	(5.5–15.5)
Cultivating		
Weeding	5.3	(3.8–7.8)
Hoeing	5.8	(4.8–6.8)
Excavating		
Digging trench, clay soil	8.5	(—)
Shoveling: 8-kg load up to 2-m lift and distance	—	(7.5–10.5)

SOURCE: R. Passmore and J.V.G.A. Durnin, "Human Energy Expenditure," pp. 821; 827–828. Data are from Europe and West Africa.

[a]Energy expenditures can be applied only to work periods of no more than 15 min. For hourly or daily expenditures time studies are necessary to account for rest and other activities.

by manipulating soil, water, topography, climate, and space—the resources that I use to organize this study. Human energy requirements for several typical tasks are suggested in Table 2-1.[9]

Energy costs are but one side of the ratio. What is gained by expending these calories? For work units as variable as humans and animals, production depends on such factors as strength, health, experience, incentives, weather, and time of day. Nevertheless, approxi-

[9]See also J. N. Black, "Energy Relations in Crop Production"; Richard C. Fluck and C. Direlle Baird, *Agricultural Energetics;* P. G. Phillips, "The Metabolic Cost of Common West African Agricultural Activities"; M.J.T. Norman, "Energy Inputs and Outputs of Subsistence Cropping Systems in the Tropics."

Table 2-2. Manual Excavation and Transfer of Regolith Materials

Activity	m^3/Man-Day[a]
Excavating[b]	
Loose, sandy loams	7–9
Moderately compacted soils and clays	3–6
Consolidated and rocky materials	0.5–3
Saturated, mucky soils	5–6
Shaping; mounding loose dirt[c]	10–24
Loading loose dirt, sand[d]	6–9
Spreading loose dirt, sand[e]	18–22

SOURCE: Field measurements by the author; daily quotas set for hired labor. See also the production rates presented in Basil Coukis, *Labor-Based Construction Programs,* and the Technical Memoranda issued by the World Bank (*World Bank Study of the Substitution of Labor and Equipment in Civil Construction*).

[a]Rates for generally healthy, experienced, reasonably motivated workers; working day usually about 6 hr.
[b]Dislodging materials and lifting about 1 m from pit or trench with picks, shovels, and spades.
[c]Shaping or mounding plowed fields with shovels or hoes.
[d]Lifting 1 to 2 m and loading into carts, wagons, or trucks with shovels or spades.
[e]Spreading from convenient piles with shovels or spades.

mations are useful. Although limited to earth and rock transfers, the data in Table 2-2 apply to a variety of resource-management tasks, such as soil management, slope control, and maintenance of gravity irrigation systems. (Lifting rates in manual irrigation systems are reviewed in a later chapter.) Some of the procedures reviewed in succeeding chapters, such as those associated with climate and space management, are less amenable to energy analysis because they involve little in the way of materials transfers and are better measured by task time and effort.[10]

In addition to managing field resources, traditional farmers perform many other operations, including seed production, crop storage and processing, maintenance of tools and buildings, transportation, and marketing. Although all these tasks require time and energy, they are beyond the scope of this study and will not be reviewed here.

[10]John K. Hatch *(The Corn Farmers of Motupe)* reports detailed task times for typical field operations.

Energy Sources

By definition traditional farming systems use little energy. The traditional farmer, however, faces exactly the same resource management challenges that the industrialized farmer does, but on a smaller scale. Lacking the industrialized farmer's fossil fuel supplements, on what energy sources does the traditional farmer depend?

Humans and Animals

Power in traditional farming comes mostly from humans and animals fueled by foods produced on the farm or in the local community. Animals are often depicted as competitors for human food or, at least, as users of space on which food for people could be grown. The description is at least partially accurate for large farms or ranches in less-developed and developed countries, where animals occupy arable lands or are fed grains and other foods that are edible by people. In traditional small-farm systems, however, animals usually use or put to higher use materials that otherwise would contribute little to human diets. Draft animals graze on crop residues, weeds, and rough pastures; pigs and chickens forage and are fed garbage.

Compared with even small fossil fuel devices, human and animal power appear puny indeed. But such comparisons are not clear-cut. Much of the power of industrial equipment is spent either moving the machine itself or on nonspecific applications. For example, tractors drawing tools expend as much energy on the bare spaces of ground between plants as they do on effective cultivation, whereas carefully guided hoes dislodge only the soil necessary to uproot weeds.

Neither animals nor tractors eliminate human labor altogether (Table 2-3). For example, farmers often complain of the difficulty of managing teams in small plots where pulled tools must be physically turned around at the end of each row. Teams sometimes are used only for heavy plowing; subsequent planting and cultivation is done by hand with less effort.

In contrast, human labor combines judgment and selection with the application of energy on a scale and with a precision otherwise impossible. For example, in animal or tractor systems a rigid weed-crop dichotomy must be established since any deviation from straight and continuous application reduces the value of the equipment. In contrast, farmers cleaning their fields in small-scale hand operations may distinguish between truly noxious weeds and those that can serve as mulch or fodder, and harvest volunteer food and medicinal plants.[11]

[11]J. C. Chacón and S. R. Gliessman, "Use of the 'Non-Weed' Concept in Traditional Tropical Agroecosystems of Southeastern Mexico."

Table 2-3. Comparative Energy Use for Maize Cultivation

	Manual Systems (10^3 kcal/ha/yr)			Draft-Animal Systems (10^3 kcal/ha/yr)[a]			Industrial Systems (10^3 kcal/ha/yr)	
	Mexico	Guatemala	Nigeria	Mexico	Guatemala	Philippines	U.S.A.	U.S.A.
Energy Source								
Labor (190 kcal/hr)	217	269	118	73	133	56	11	11
Draft animals (team) (190 kcal/hr)	—	—	—	376	591	277	—	—
Tools and machinery	11	11	11	242	242	242	444	1,037
Fuels and electricity: field equipment, irrigating, drying	—	—	—	—	—	—	1,640	3,379
Chemicals: fertilizers, pesticides, herbicides	—	—	215	—	—	90	190	2,708
Total inputs	228	280	344	691	966	665	2,285	7,135
Outputs: maize (3,600 kcal/kg)	6,962	3,802	3,578	3,352	3,802	3,352	7,636	18,216
Energy ratio: edible energy output ÷ energy input	30.5	13.6	10.4	4.9	3.9	5.0	3.3	2.6

SOURCE: Data modified from Leach, *Energy and Food Production,* pp. 134–37. See also Michael R. Goe and Robert E. McDowell, *Animal Traction.*

[a]Although human labor is reduced in animal systems, overall efficiency is not notably increased. Black, "Energy Relations in Crop Production."

Such differences muddle comparisons of efficiency between human labor and other sources of power. Energy comparisons of tasks involving large-scale mass transfers, such as field leveling, are valid and usually favor mechanical equipment. But for delicate tasks such as planting and cultivation where human dexterity and ingenuity focus and conserve energy expenditures, comparisons are of questionable value and may be misleading.

Another reason it is misleading to compare energy efficiency in agriculture concerns specific objectives. Certainly all farmers are interested in the same or greater returns for less effort. In the extreme example of subsistence agriculture, overall energy returns must exceed expenditures or the systems will fail. But efficiency has never been the sole individual or societal goal. Even in subsistence systems particular tasks may result in net losses but are justified by such nonquantifiable returns as security or prestige. On a larger scale, diet can be viewed as the basic objective of the economic organization of society.[12] From this perspective, food is both a means and an end of production; this is not true for other forms of energy.

The limited power of individual humans or animals is often insufficient for tasks that involve substantial mass transfers or tight schedules. A large boulder or log, for example, may not be reducible to a size that can be managed by one or two people or a family. Similarly, planting before the onset of rains imposes an external deadline that human- or animal-powered systems may have difficulty meeting. In such cases the advantages of massed or collective effort are apparent. Total man-hours or energy expended in collective effort may actually be greater than if individuals performed the tasks alone. Interference and time spent coordinating and supervising often affect efficiency adversely.[13] But psychological benefits, particularly on arduous or tedious jobs, may outweigh these disadvantages.

Resource management that requires the combined efforts of families, villages, or regions exerts a strong influence on the structure of the practicing societies. This is especially true of water management, which in many traditional societies is a primary force in social and political organization.

[12]E. H. Ashton, "A Sociological Sketch of Sotho Diet."

[13]One of the most common ways to combine effort is to join two or more animals in teams. But here too efficiency declines as more animals are harnessed together. A team of six animals, for example, is less than two-thirds as efficient as a single animal. World Bank, *World Bank Study of the Substitution of Labor and Equipment in Civil Construction,* Tech. Mem. 21, "A Literature Review of the Work Output of Animals," p. 10.

Natural Forces and Sources of Energy

In addition to humans and animals, nonindustrial farmers have access to other sources of energy for resource management.

Gravity

Overcoming gravity is the single greatest energy sink in traditional resource management. But this ubiquitous force can also be used to advantage. For example, the masses of water used in irrigation and drainage in most traditional systems move through delivery networks constructed with slight gradients. The key is control; gravity is an accelerating force, and down-slope transfers can quickly become destructive if speeds are too high.

Gravity can also assist with other tasks, such as down-slope plowing (to the despair of conservationists, who much prefer contour cultivation) and (with or without water) terrace construction. Management methods that work with rather than against gravity are discussed in the chapters on particular resources.

Fire and Vegetation

Fire is one of the most recognized yet least analyzed sources of energy in traditional farming. It is used primarily to dispose of organic debris in shifting cultivation systems, to clear plots of crop residues, to destroy weeds and weed seeds, and to produce fertilizing ash. It is, as Stewart remarks, "the first great force employed by man" and remains a faithful friend of traditional farmers the world over.[14]

But unlike the standard fossil fuels used in industrialized agriculture, the fuels of traditional fires usually are local, diverse, often unwanted vegetable materials whose caloric value and market price can only be estimated. Furthermore, the caloric costs of such fuels customarily are not included in energy balance sheets, which gives the traditional cultivators an unfair advantage over their counterparts who burn fossil fuels.[15]

In addition to conveniently destroying themselves, living and dead plants play a role in the management of other resources, such as water (stake dams, conduits), space (fences, trellises), and climate (shade,

[14]Omer C. Stewart, "Fire as the First Great Force Employed by Man"; H. H. Bartlett, "Fire, Primitive Agriculture, and Grazing in the Tropics."

[15]M.J.T. Norman (*Annual Cropping Systems in the Tropics*, p. 125) calculates that burning a secondary forest with standing dry matter of 100 tons ha consumes 4.8×10^8 kcal of energy. This amount of energy is several orders of magnitude greater than all other energy inputs combined in shifting cultivation.

windbreaks). But these are passive, not active, sources of energy for resource management.

Wind and Water Currents

The energy in wind and streams has been tapped by traditional farmers elsewhere, especially in Europe and colonial North America, where various types of mills were used to lift water for irrigation and drainage and to grind grain. But wheels and other devices to convert lineal to rotary motion, and vice versa, were never developed in the pre-European New World, and wind- and water-driven mills have not been widely adopted by the small-farm sector of Middle America.[16]

Capillary Transfers

Capillarity is not usually included in a list of energy sources for on-farm resource management. But capillary action performs the "work" of lifting water in several important Middle American irrigation systems and thus deserves a place in this inventory. Capillary transfers are examined in detail in Chapter 10.

Fossil Fuels

Limited use of fossil fuels is a distinguishing feature of small farms and the most common explanation for their low labor productivity. Certainly most traditional farmers are well aware of the potential of these forms of energy and would quickly employ them given the opportunity.

The two general areas where fossil fuels offer the greatest immediate benefits (as does combining the efforts of humans or animals) are mass transfers and scheduling. Horizontal movements of soil and vertical lifting of water are among the most arduous tasks faced by traditional farmers. Little wonder that trucks and pumps often are the first powered tools they adopt. Similarly, labor and draft-animal shortages at critical plowing or planting times encourage the use of tractors.

But the fuels themselves, the engines and motors needed to convert them to power, and the tools by which that power is applied to resource management involve capital and operating costs well beyond the financial capacity of most traditional systems. In addition, fossil fuels and equipment originate outside the local region, as does the operation and maintenance technology. These obstacles to adoption are well known and need no elaboration here, especially because the focus of this study is on resource management without recourse to fossil fuels.

[16]Gene C. Wilken, "Manual Irrigation in Middle America."

Other Strategies

Scale and Scheduling

Another response to the power limitations of traditional agriculture is to reduce scale, to divide tasks into spatially and temporally manageable units. Traditional holdings typically are small, matching in size the work capabilities of humans and animals. However, the relationship between space and work is not a simple one because farm size is only one indicator of work requirements.

Within the farm, further reductions are possible: small individual plots that can be plowed, planted, cultivated, and harvested in one or a few days; crop and row spacing suitable for hoe or plow cultivation; canals that can be maintained with a shovel. For tasks that are not easily divided, group or collective effort may be substituted for individual labor, as I discussed earlier.

Small scale also facilitates adjustments to environmental hazards. For example, tiny seed beds can be protected against hail, heavy rain, and birds with organic or plastic covers. Mulches and windbreaks shield small plots from frost and lodging. Hand-irrigated fields are small enough to be watered every day should a drought require such frequent application.

Management as Process

Still another strategy employed in traditional resource management might be called patience and persistence, in which the energy required for sizable tasks is applied in small increments at irregular intervals. Tight schedules and target dates are a luxury of high-energy, industrialized systems, where the task itself and external factors determine the type and size of equipment that will be used and thus the rate at which energy will be expended. Lacking concentrated power sources or the facilities with which to acquire them, traditional farmers invest labor only as it can be spared from routine farm operations to achieve changes over long periods of time. In this sense many traditional resource-management activities are ongoing processes more than discrete projects.

The seasonal march of moisture and temperature dictates annual cropping schedules and associated resource management. But lengthier efforts, such as terracing (slope management) or irrigation works (water management), may be undertaken with little thought as to pace of work or final completion date. To Western eyes the procedure is untidy; half-terraced hillsides, partially constructed canals, and unfinished roads suggest a lack of industry and planning. But often these are projects under way; they represent a process of cautious capital formation analogous to a small savings program, where excess labor instead of money is banked.[17]

Invested Labor

Resource management requires periodic and ongoing energy inputs for ecosystem modification and maintenance. Because biophysical processes do not vary, traditional farmers practice the same sorts of resource management as do industrial farmers. But, without recourse to fossil fuels, traditional farmers personally do the work. Thus their principal investment is not in machines but in end results, mostly the reshaping of surface topography (terraces, field leveling, canals, drains)—an investment not in capital but in labor.[18]

The two investments are, however, quite similar. Labor is "saved" from farm and leisure time, "banked" bit by bit in a resource-management project, and eventually "returned" in the form of reduced operating costs (less labor) or higher yields. Although it should be possible to calculate rates of returns on invested labor, lack of exact factor pricing makes the job difficult.

The analogy can be extended even to credit; it is possible to "borrow" labor through cooperative work arrangements. But these arrangements are commonly associated with routine field operations such as planting and harvesting, where seasonal climate regimes impose rigid schedules on task completion. Outside labor for long-term resource-management projects is customarily paid in cash.

Information Systems

Information systems were discussed briefly in Chapter 1. Their role in traditional management is particularly important because they include the means by which scarce energy is focused and applied. By word and example, farmers convey to children and peers information on how to make individual efforts efficient.[19]

Tools

This book contains little discussion of equipment. For the most part traditional tools are general-purpose, simple, and familiar to

[17]In this case "excess labor" does not mean labor in excess of overall agricultural needs but rather labor taken from slack time, leisure, or nonfarm activities and applied to constructing permanent structures and facilities.

[18]Martin L. Primack ("Farm Capital Formation as a Use of Farm Labor in the United States, 1850–1920") discusses the concept of invested labor in agriculture.

[19]More than one writer has commented on the economy of motion that can be achieved in repetitive farm tasks (for example, Hatch, *Corn Farmers*).

farmers and readers everywhere, and only brief comments are needed to clarify local design or use. A few exceptions are so specialized (for example, Almolongan water shovels; see Chapter 9) or complicated (for example, Tehuacán horizontal wells; see Chapter 10) that they merit extended description. But generally, traditional technology consists of ways of doing things rather than equipment for doing them.

Summary

Reliance on local energy sources and human or animal power is often considered a major deterrent to development. It is held that even if, by strenuous efforts, high yields are achieved, lack of fossil-fuel subsidies condemns small farmers to low productivity per unit of labor. In the case of human labor, the ultimate obstacle is energetic equivalence. Unless a system yields substantially more calories than are consumed in its operation, it is doomed to failure.[20]

But we have seen that sources of energy other than human, animal, or fossil fuel, and economies of energy application make this sort of easy analysis suspect. Lacking power to offset natural forces, traditional farmers instead put them to work. Rather than using too much power on farm tasks, as is common in energy-rich societies, traditional farmers carefully focus the limited power at their disposal in manageable units at critical points. This cautious marshaling of scarce resources is largely responsible for the favorable energy input/output ratios that have come to characterize traditional agricultural systems.

[20]Industrial systems consistently exceed energetic equivalence by exchanging cheap fossil fuel for expensive food calories. David Pimental et al., "Food Production and the Energy Crisis." But see E. C. Pasour, Jr., and J. Bruce Bullock, "Energy and Agriculture," for a critique of energy budget studies from an economic perspective.

3

Soil Classification

Soil management provides an excellent introduction to traditional resource management because it illustrates most of the common elements: practices ranging from simple to complex, use of several types of energy, materials obtained from local sources, and detailed classification systems. It also reveals the deep understanding of resource characteristics and of requirements for sustained production that farmers have in the best traditional systems.

Yet, despite the importance of this resource, few researchers have studied traditional soil use or management. Those who study soils ignore achievements in managing soils, while those who study management ignore agronomy; we are left with the impression, entirely incorrect, that soil science is unique to technically advanced societies.

Soil classification offers insights into the perspectives of farmers and suggests some of the management options open to them. Not surprisingly, traditional classification systems tend to be utilitarian rather than taxonomic.[1] Because soils are used in a variety of ways in traditional societies, many groupings are possible. Agricultural classifications of soil are foremost, but, in addition, pottery, construction, and other soil-using industries have developed categories for soil.[2]

[1]Nyle C. Brady (*The Nature and Properties of Soils,* pp. 317–22) contrasts the pedological approach, which uses genesis and intrinsic properties for classification, with the edaphological emphasis on soil properties in relation to plants.

[2]For example, see Dean E. Arnold, "Ethnomineralogy of Ticul, Yucatan Potters."

Agricultural soil classification is an ancient practice in Middle America.[3] Sahagún provides glimpses of the complexity that had developed even in precolumbian times.[4] The two dozen or so agricultural soils that he describes are identified by origin or source (sedimentary, bog, leached, hill), color, texture, smell, consistency (soft, spongy, greasy), and organic content. These characteristics are related in turn to agricultural evaluations: fertile, productive, workable (perhaps only when wet), or useless.[5]

The evaluations reported by Sahagún emphasize inherent qualities and tend toward absolutes; for the most part soils are either naturally good or worthless. But there are a few examples of soils whose value could be enhanced by treatment. For example:

> *Tlalauiyac:* This is the land [soil] which is good, which produces, which is mellow. I fertilize it. I add humus to it. I make it mellow. I make it good.[6]

Although fragmentary, such evidence suggests that soil management, like soil classification, was well advanced in early times.

Only a few studies reveal contemporary systems of comparable complexity. For example, Kekchi farmers in the Guatemalan lowlands near Lake Izabal use four criteria to identify and evaluate soils: color (eight categories), texture (ten categories), drainage (five categories), and root content (one category, root bound). Evaluations depend on combinations of these sometimes contradictory qualities.[7] The Kekchi consider soil of primary importance in crop production. But they are shifting, not sedentary, cultivators and lack a technology for improving soil quality beyond the incidental addition of ash from burned vegetation.

To demonstrate the principles of traditional soil classification, the systems currently in use in two Middle American *municipios* (approximately, counties) are reviewed in this chapter. These administrative units are convenient because they are large enough to be reported separately in national agricultural and population censuses, yet small enough to be culturally homogeneous; traditional soil science and terminology tend to be relatively uniform throughout. In addition, the

[3]Rayfred L. Stevens, "The Soils of Middle America and Their Relation to the Indian Peoples and Cultures," especially pp. 267, 303–4.

[4]Bernardino de Sahagún, *General History of the Things of New Spain,* Pt. 12, pp. 251ff.

[5]For analyses of Aztec soil classes and their continuity in a modern Mexican community, see Barbara J. Williams, "Aztec Soil Science"; "Middle American Folk Soil Taxonomy"; and "Pictorial Representation of Soils in the Valley of Mexico."

[6]Sahagún, *General History,* Pt. 12, p. 252.

[7]William E. Carter, *New Lands and Old Traditions,* pp. 20ff.

lands of a *municipio,* or even of an individual village, may include several of the region's general soil types.

The review of soil classification here consists of three steps: an overview of soil conditions as revealed by modern soil surveys; a listing of local soil types and their characteristics as identified by farmers; and laboratory analyses of soil samples for comparison of physical and chemical properties with sensible (perceived) descriptive differences.

Communication between those who conduct official soil surveys and those who till the soil is seldom good, and few traditional farmers are aware that official classifications and analyses exist. Instead, each physical or ethnic unit has its own system for identifying types of soils and evaluating them for agricultural purposes. In most cases, however, traditional and scientific appraisals correspond rather closely, as would be expected. What distinguishes local classification from regional reconnaissance is degree of detail; farmers recognize subtle variations and combinations of soils overlooked by most surveys. Even tiny plots may have several distinct types of soils that farmers identify and for which they adjust cropping and management practices.

Thus, each generalized soil map in this chapter has superimposed on it in actuality a much more detailed map of local classifications. If these local classifications were included, they would provide a mosaic of great detail and complexity—the actual soil resources of individual farmers—which would be of value in understanding local techniques of crop and soil management.

We now turn to an inspection of soil classification in the two *municipios*: Tetlatlauhca, in the state of Tlaxcala, Mexico; and Almolonga, in the department of Quezaltenango, Guatemala. In both areas cultivation is intensive. In addition to subsistence maize, irrigated fields are planted to a variety of vegetables destined for local and regional markets. Soil technology is traditional; classification, evaluation, and management methods are products of local experience, and most amendments for maintenance and improvement of soil fertility are products of the immediate region. The farmers in both *municipios* also are knowledgeable about chemical fertilizers and apply them as needed or when they can afford them. But the adoption of this exogenous technology has taken place within an existing traditional system of considerable sophistication.

Tetlatlauhca, Mexico

The southwestern part of the state of Tlaxcala is a region of nearly flat basin-floor lands at about 2,200 m elevation, rimmed and broken by low hills of andesites, tuffs, and breccias. To the east towers the extinct volcanic peak of La Malinche (4,461 m), fifth highest mountain in Mexico. Like most of the basins on the Mesa Central, the Puebla-Tlaxcala basin held a large lake in Pleistocene times.[8] Vestiges remain in the form of swamps and bogs, and the process of filling continues. Soils on the basin floor are mostly alluvium from adjacent highlands.

The region is densely populated, with an average of more than 200 persons/km^2 (500/mile^2).[9] Most people live in some form of community, usually nucleated villages *(pueblos)* that are characteristically located at the break between hill slope and flat lands, where fertile alluvial or lacustrine soils are easily accessible but not wasted under buildings. Many farmers own or operate[10] both hill and bottom lands and must be conversant with a variety of soil types (Figure 3-1).

Climatically, the region is semiarid, with pronounced wet and dry seasons. More than 70 percent of the 800-mm annual precipitation falls during the five months from May through September. Average monthly temperatures vary but little, ranging from December lows of 20°C to 23°C to highs of 25°C to 28°C in April, when increasing cloudiness truncates the rising curve of annual temperature. Winter frosts are not uncommon and can be severe, although the ground never freezes.

Two modern soils surveys of the region are available: a statewide survey by Cruz López, which forms part of a general agronomic study;[11] and the more recent surveys conducted during the course of the Proyecto Puebla-Tlaxcala.[12] Few traditional farming regions can boast such coverage. The review here of soils in southwestern Tlaxcala is taken from these surveys.

Deep sands of the Malinche series, derived from volcanic sands, ashes, and andesites, appear on the wooded upper slopes of La Malinche, well above hill and basin-floor agricultural soils. But these structureless sands are easily eroded as soon as protective vegetation is removed, and it is primarily Malinche-series sands that choke the

[8]Luis Bláquez L. and Raúl Lozano García, "Hidrogeología y Minerales No-metalicos del Estado de Tlaxcala," p. 22.

[9]Dirección General de Estadística (Mexico), *IX Censo General de Población, 1970,* Tables 1, 25.

[10]In addition to owning, leasing, or sharecropping land, farmers may also operate *ejidos,* the village-administered lands expropriated from haciendas under the Mexican land reform. A few parcels of communal land, also administered by the villages, generally are of poor quality and are used only for grazing.

[11]Candido Cruz López, "Estudio Agrológico Regional del Estado de Tlaxcala," No. 3, pp. 81–117.

[12]Hans Aeppli, "Aclaraciones acerca del Diseño del Mapa de los Suelos de la Cuenca de Puebla-Tlaxcala"; Gerd Werner et al., "Los Suelos de la Cuenca Alta de Puebla-Tlaxcala y sus Alrededores."

Figure 3-1. Landscape in southwestern Tlaxcala, Mexico. View from low hills across basin floor to lower slopes of La Malinche.

barrancas (ephemeral streamways) and spread out over cultivated fields during floods. Although not necessarily infertile, the sands lack organic matter, and fields on which they have been deposited must be treated with manure before cultivation.

Thin sands and sandy loams of the Tlaxcala series are found on the slopes of La Malinche and, together with similar alluvial- and tuff-derived soils, on the slopes and crests of adjacent highlands to the north and on isolated hills on the flood plain. Tlaxcala-series soils are weathered in place from consolidated tuff, or *tepetate*. Generally, they are rich in minerals, although superimposition of sand deposits may reduce their fertility. These soils are extremely sensitive to sheet and gully erosion. The uppermost, or A, horizon (soil layer) is frequently missing, and in places even the next, or B, horizon has been stripped off, exposing great stretches of bare *tepetate* and contributing to the region's dubious distinction of being the worst eroded in Mexico.[13]

Tepetate remains something of an enigma. The name comes from the Nahuatl *tetl* (rock) and *petatl* (mat), literally and rather accurately a

[13]Cruz López, "Estudio Agrológico," No. 1, p. 57; No. 3, pp. 91–94; Jorge L. Tamayo, *Geografía General de México,* vol. 3, p. 49.

mat of rock.[14] Often it is equated with calcium deposits *(caliche)* or other mineral pans that form when horizons develop concentrations of precipitates.[15] But Mexican geologists and agronomists accurately define *tepetate* as tuffacious material resulting from geological processes of deposition and consolidation.[16] Wells drilled in southwestern Tlaxcala reveal multiple layers of *tepetate* to depths below 100 m.[17]

Poorly structured Zacatelco-series sands and sandy loams form a transition between basin-floor and upland soils. Originating almost entirely from volcanic andesites, these deep soils were alluvially deposited on the lower slopes of La Malinche, on lesser hills, and on the basin floor. Drainage is good except on the plain near Natívitas-series soils, where water tables are shallow.

The most extensive flood-plain soils are the sandy to clayey loams of the Natívitas series. These soils are deep, badly drained, and occasionally spotted with salt accumulations *(tequesquite)*. Ground-water levels often are near the surface. Soil nutrients are good, however, so agriculturally the basic disadvantage is high water tables.[18]

Loams and sandy loams of tuff and andesite parent material and of ordinary fertility are the second most extensive alluvial soils on the basin floor. These soils, mapped as the Muñoz series, often rest on sedimentary deposits of consolidated *tepetate*, although it is uncertain whether they do so in midbasin areas, where Muñoz-series soils tend to follow river courses. Internal drainage is good on high ground but poor in the lowlands.

The *municipio* of Tetlatlauhca in southwestern Tlaxcala has a total area of 44.6 km^2 (4,460 ha), of which 2,582 ha are classified as productive. Of these, 2,459 ha, or 95.2 percent, are workable lands *(tierras de labor)*, with the remainder made up of pasture, unworked lands, and so forth (Table 3-1). Although most farm land is rainfall irrigated *(temporal)*, the *municipio* also contains 584 ha of irrigated lands *(riego)* and 284 ha of subirrigated (*jugo* or *humedad*, literally, juicy or moist) lands.[19]

[14]Although the population is Spanish speaking, indigenous (Nahuatl, or Mexicana) terms persist for many elements in the rural environment, including soils.

[15]For a review of the confusing literature, see Barbara J. Williams, "Tepetate in the Valley of Mexico."

[16]Miguel R. Brambila, "The Tepetate Soils of Mexico"; Cruz López, "Estudio Agrológico," no. 3, p. 83; Alfonso García Espinoza, "Estudio de los Suelos Tepetatosos y las Posibilidades de su Recuperación Agrícola."

[17]Secretaría de Recursos Hidráulicos (Mexico), "Obras en el Estado de Tlaxcala."

[18]Gene C. Wilken, "Drained-Field Agriculture."

[19]Dirección General de Estadística (Mexico), *IV Censo Agrícola, Ganadero, y Ejidal, 1960,* Table 1, p. 8.

Table 3-1. Classification of Productive Lands, Tetlatlauhca

	ha	%
Total productive lands	2,581	100.0
Workable lands (*laborable*)		
Rainfall-irrigated (*temporal*)	1,587	61.5
Irrigated (*riego*)	584	22.6
Subirrigated (*jugo* or *humedad*)	284	11.0
Orchards (includes agaves)	4	0.2
	2,459	95.3
Pastures	22	0.9
Forests	22	0.9
Uncultivated	78	3.0
	122	4.7

SOURCE: Dirección General de Estadística (Mexico), *IV Censo Agrícola*, Table 1, p. 8.

The *municipio* is densely populated and overwhelmingly agrarian. With a total population in 1970 of 10,208, gross density was 229/km^2 (585/mile^2). Of the 2,622 persons twelve years of age or older who were economically active, 1,971, or 75 percent, were engaged in agriculture.[20] The average workable land area per person economically active in agriculture was thus about 1.25 ha (2,459 ÷ 1,971).

Many of the soil types noted for southwestern Tlaxcala occur within the *municipio* of Tetlatlauhca (Map 3-1). In addition, local conditions, such as slopes, high water tables, and poor drainage, create a variety of microedaphic situations. Because many farmers have plots both on the basin floor and in the hills, they must learn to cope with a wide range of soil conditions.

Traditional soil identification and evaluation are based on a limited number of sensible characteristics, primarily color and texture. Other factors, such as structure, salinity, depth, and parent material (for example, *tepetate*) are included if important. Shadings of major soil characteristics are indicated by modifying descriptive terms (reddish yellow, slightly salty). Soils are rated superior or inferior primarily on the basis of anticipated crop yields. Yet, even with this almost limitless array of descriptive possibilities, farmers are reluctant to pass final

[20]Dirección General de Estadística (Mexico), *IX Censo General de Población*, Tables 1, 25.

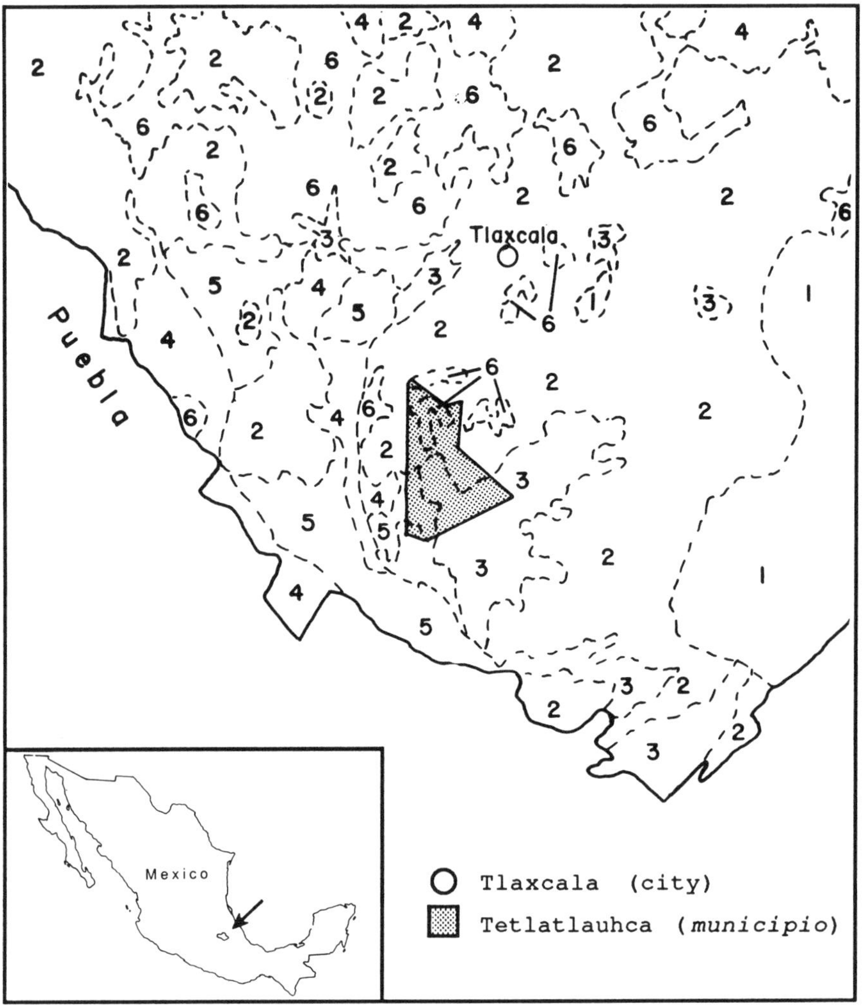

Map 3-1. Soils of southwestern Tlaxcala.
SOURCE: After Cruz López.

1. Malinche-series volcanic sands.
2. Tlaxcala-series sands and sandy loams; in highlands often a thin, complex soil overlying *tepetate*.
3. Zacatelco-series piedmont sands and sandy loams.
4. Muñoz-series flood-plain loams and sandy loams.
5. Natívitas-series flood-plain sandy to clayey loams.
6. Exposed *tepetate* and thin hill soils overlying *tepetate*.

judgment on a particular soil from inspection only. They feel rather that a soil's true nature can be known only by working with it over several seasons.

Different soils require different management practices. For example, some soils can be planted to any of the common crops (maize, beans, onions), whereas others are good for only one or two crops or for a specialty crop. Fields with soils that are workable only when wet cannot be prepared until the onset of soaking rains, usually in late May or June. Thus, constraints are imposed on planting and, because of crop maturation periods (growing seasons), on the types or varieties of crops that can be planted. Because poor *(pobre)* or weak *(debil, flojo)* soils support fewer plants, plant spacing is in part a result of soil quality. In some soils, fertility is markedly enhanced by dressings of animal manure or chemical fertilizer; in others, it is not. Soil management and amendments are treated at length in Chapters 4 and 5, and are mentioned here only as they relate to soil classifications and evaluations.

Eventually, soil quality, related environmental conditions (water, slope), and location determine land prices.[21] In Tetlatlauhca, basin-floor plots with first-class soils (for example, *tierra fuerte,* compact soil) brought 6 to 9 pesos (US$0.50 to $0.75) (or more)/m^2 (US$5,000 to $7,500/ha) in 1974, whereas hill fields with thin soils might have sold for a fifth of that price or less.[22]

The following list of soils, compiled from farmers' responses, suggests the range of soils and bases for identification and evaluation. As with soils anywhere, two farmers might disagree on particular details. But generally soils and their qualities are readily recognized by most farmers in the region. A brief description paraphrased from farmers' responses accompanies each listing of soil type. Chemical and mechanical analyses are given in Table 3-2.[23]

Hill Soils

Hill soils have developed in place on the andesites, tuffs, and breccias that make up the highlands of the region, and on the various

[21]The volume of transactions within a traditional village is not enough to create a land market or a price structure. Farmers' estimates of relative values suggest that land prices are strongly correlated with soil quality.

[22]Despite low income levels, land is not necessarily inexpensive. Prices for good irrigated farm land as high as US$1.00/m^2 (US$10,000/ha) were reported in southwestern Tlaxcala. In a densely populated part of Guatemala, one study noted prices of US$2,300/0.1 ha. Escuela Facultativa de C. C. Económicas de Occidente y Comité Interamericano de Desarrollo Agrícola, *Tenencia de la Tierra y Desarrollo Socio-Económico del Sector Agrícola en Guatemala,* pp. 177–81.

[23]All soil analyses by the Soil Testing Laboratory, Colorado State University, Fort Collins.

Table 3-2. Properties of Farmer-Identified Soils, Tetlatlauhca

	pH	Cond.	O.M. (%)	NO_3-N (ppm)	P (ppm)	K (ppm)	Zn (ppm)	Fe (ppm)	Lime (%)	Sand (%)	Silt (%)	Clay (%)	Texture	meg/ 100 g CEC
Hill soils														
Barro amarillo	6.2	0.3	1.0	2	2	260	0.6	13.4	Low	42	22	36	Clay loam	21.8
Barro blanco	6.4	0.3	0.5	1	1	+500	0.3	3.0	Low	23	43	34	Clay loam	69.0
Barro colorado	6.8	0.3	0.4	2	1	465	0.3	7.6	Low	40	29	31	Clay loam	39.0
Barro pardo	6.2	0.6	2.0	9	17	380	0.7	20.7	Low	43	23	34	Clay loam	11.2
Tierra blanca	8.0	0.5	1.9	3	6	313	0.6	2.8	High	52	12	36	Sandy clay loam	18.4
Tepetate soils														
Tepetate amarillo	7.5	0.6	0.2	0.8	1.3	398	0.5	9.6						
Soil derived from *tepetate amarillo* after first year of cultivation	6.5	0.4	1.0	0.4	0.5	203	0.6	17.0						
Soil derived from *tepetate amarillo* after several years of cultivation	7.5	0.7	1.0	8.0	1.3	440	0.3	11.4						
Tepetate colorado	6.8	0.2	0.4	1.2	3.5	423	0.4	11.4						
Soil derived from *tepetate pardo* after several years of cultivation	6.9	1.2	0.3	11.6	1.0	293	0.6	23.6						
Basin-floor soils														
Tierra fuerte	7.2	0.6	5.3	6	13	484	0.9	+40.0	Low	19	42	39	Silty clay loam	39.8
Tierra arenosa	8.1	0.6	1.6	3	1	258	0.3	12.2	Medium	64	12	24	Sandy clay loam	19.8
Tierra negra	6.0	0.8	6.7	5	5	+500	2.0	+40.0	Low	—	—	—	—	34.0
Salty soils														
Tequesquitudo	8.3	0.6	1.1	1	1	195	0.3	8.9	Medium	78	7	15	Sandy loam	11.8

sands, ashes, and clays laid down during periods of vulcanism and alluviation.[24] Generally, hill soils are easily eroded. Because population densities and pressures on vegetation have been high for centuries,[25] soils on most hills are seriously depleted. Deep *barrancas,* thin soils, and exposed *tepetate* characterize the uplands of southwestern Tlaxcala.

Barro amarillo (yellow clay) is a good but not exceptionally fertile clay loam; it is sticky *(fuerte)* and difficult to work. Applications of manure improve the fertility of *barro amarillo* and also make it easier to plow. Dressings of sand also would improve texture, but transporting sand from *barranca* sources to hill fields is so arduous with animals or expensive with trucks that it is beyond the capacity of most traditional farmers.

Barro blanco (white clay) is a whitish clay loam of indifferent quality derived from light-colored clay deposits. Fertility and tilth can be improved with manure, but often available amendments are saved for more valuable basin-floor soils.

Barro colorado (red clay) is a sticky *(fuerte)* clay soil of mediocre quality. Generous applications of manure are needed to improve fertility and tilth, but these are only infrequently made.

Barro pardo (dark clay) is an exceptionally tough *(fuerte)* clay loam that can be worked only when wet. Because field preparations must be delayed until the onset of soaking rains, only quick-maturing maize varieties or other short-season crops can be planted on *pardo* clays. Thus, the generally poor rating of this soil is based on both low fertility and management restrictions.

Tierra blanca (white soil) is a light, sandy clay loam with high lime content derived from *jis,* a soft *(bofo),* white, calcareous rock. Because *tierra blanca* is of low quality, only a few crops, such as onions and husk tomatoes, do reasonably well on it. Maize plants remain small and yellow even with application of manure, possibly because of the low iron content.

Outcroppings of white calcareous rock are not uncommon in the hills of southwestern Tlaxcala. *Jis,* the soft, white rock on which *tierra blanca* develops, has been mentioned. Relatively harder *cal* (lime) is excavated, pulverized, and dissolved in the water in which maize is soaked prior to grinding for tortillas.

Tepetate Soils

Tepetate soils differ from those already noted as developing on *tepetate* (for example, Tlaxcala-series sands and sandy loams) in that

[24]Blásquez L. and Lozano García, "Hidrogeología y Minerales No-metalicos," pp. 13ff.

[25]Wilken, "Drained-Field Agriculture."

they are recent, with no or poorly developed profiles. In large part, they are human-made soils (anthrosols or ethnosols) created in the following fashion. Where soil depth is less than plow depth, a few centimeters of *tepetate* are incorporated into the overlying soil each season with normal plowing. Where erosion has reduced the soil layer to less than planting depth or to exposed *tepetate,* farmers may break up and pulverize the *tepetate* with plows, picks, crowbars, sledgehammers, or even black powder, to form a usable soil layer. Although lacking in organic matter, some kinds of *tepetate* have the chemical and physical properties to produce a reasonably good soil.

Tepetate amarillo (yellow) is a light, weakly consolidated formation that is relatively easy to break up. The resulting soil has good working properties (not *fuerte*) and, after organic materials have been added, is of fair quality. (The high NO_3-N value in Sample B-1, Table 3-2, is due to the addition of chemical fertilizers.)

Tepetate colorado (red) is a medium hard, reddish *tepetate.* Although much less preferred than *tepetate amarillo, tepetate colorado* is acceptable after additions of nutrients and organic matter.

Tepetate pardo (dark) is well cemented and resists pulverizing. *Pardo* soil developed from it tends to be cloddy and difficult to work. *Tepetate pardo* is seldom used for agricultural soil, and those areas where it has been exposed are often not used for farming. The *pardo* soil shown in Table 3-2 (B-3) has been under cultivation for several years and shows the effects of amendments, especially nitrogen.

A number of other *tepetates* appear in southwestern Tlaxcala, including *gris* (gray), another hard, seldom-used tepetate, and *xalnene,* a light-colored formation sometimes quarried for building blocks.

Basin-Floor Soils

Basin-floor soils are depositional, ranging from heavy, black bog soils *(tierras negras)* to lighter loams, which may include volcanic sands washed down from the crest of La Malinche. Most are mixtures of soils developed in Pleistocene lacustrine deposits and in more recent wind and water deposits from adjacent and distant highlands.

Tierra fuerte (compact or sticky soil) is sometimes called *tesoquite* (from the Nahuatl *tetl,* rock, and *zoquitl,* clay). These are silty clay loams of superior quality. Although difficult to work because they stick to plow blades (from whence the name *fuerte*), farmers rate these soils highly and expect continuously good maize crops from them. Soil fertility is maintained with animal manure or chemical fertilizers or both, and texture may be improved by adding sand.

Tierra arenosa (sandy soil) is a sandy loam of good quality although generally ranked below *tierra fuerte.* It is easily worked whether wet or dry. Sandy soils are further distinguished by grain size. *Tierra arenosa gruesa* has large sand grains and pebbles, is least fertile, and does not respond well to dressings of manure. Sands in *tierra arenosa delgada* are finer and hold water reasonably well. Yields from *delgada* soils improve noticeably after generous applications of manure and chemical fertilizer.

Tierra negra (black soil) is a high organic-content soil from reclaimed bog lands. In recently drained fields, remaining root mats of *tule* (*Scirpus* spp.) and other swamp plants impede plowing. Quality is rated good to excellent, however, especially after a few years of cultivation.

Salty Soil

Tequesquitudo, sometimes called *tequixquitlalli* (from the Nahuatl *tequixquitl,* salt or sweat, and *tlalli,* earth), is any soil, often a *tierra arenosa,* with high salt content. In some cases salts form a visible crust on the surface. Although low concentrations may not seriously impede crop development, farmers view any signs of salt with apprehension. High salt concentrations may result in abandonment of a plot to pasture or, alternatively, may require expensive treatment.

Traditional Nomenclature and Modern Classification

In these main traditional classifications of soil in Tetlatlauhca, descriptions and properties may be joined in the name (for example, *tierra fuerte*). In any event, descriptions and properties are always linked in the minds of farmers, who classify and evaluate on the basis of both inherent and management characteristics. Thus, the *tierras arenosas delgadas* are generally considered good soils both because of their agricultural qualities and because of their workability and response to fertilizers. In this sense, classifications of managed soils differ from those used in less intensive systems (for example, shifting cultivation), where only inherent characteristics are important.

Although only major traditional soil classes have been discussed, farmers obviously can modify descriptive adjectives and nouns to create an enormous lexicon of soil terms. For example, if a farmer were to describe the soil on his plot as *"tierra fuerte, pero poco arenosa,"* his listener would immediately know that generally it was a high-quality but sticky clay loam meliorated by more sand content than is normal on *fuerte* soils. The *fuerte* soil also could be somewhat more sandy *(algo arenosa)* or even quite sandy *(bastante arenosa),* until it grades into an *arenosa* soil with relatively high clay content. Similarly a hillside *"barro amarillo poco colorado*" indicates a slightly reddish but basically yellow soil. *Tepetate* soils vary depending on the amount of original soil and organic matter.

Thus, although traditional Tlaxcala soil classification establishes only a few major types, it is almost infinitely flexible, permitting modifications that include even slight variations. Because descriptive terms, such as color and texture, relate to both fertility and manageability, soil properties can be identified with considerable precision.

Through extensive governmental and scientific surveys and through local experience, the soils of Tlaxcala are quite well known. But little evidence indicates that the information generated by modern studies has entered the local technology. There is even less evidence that detailed traditional knowledge was considered in the surveys.[26] The two classification systems cover the same agricultural space but occupy different different cultural spaces. This cultural barrier, like a layer of dense *tepetate,* permits little exchange of information.

Almolonga, Guatemala

Highland Quezaltenango is notorious for its precipitous landscapes. Still-active volcanos and frequent earthquakes have created a jumble of mountains, hills, and steep escarpments cut by deep canyons and narrow valleys. The abrupt topography assures that climatic and edaphic conditions will vary markedly over even short distances. Most of the soils are derived from volcanic materials. Pumice filling of large intermontaine basins has provided comparatively level land in the Quezaltenango and San Marcos valleys.[27] Flat, alluviated areas are found only along a few streams.

The department of Quezaltenango has an overall population density of about 143 persons/km^2 (372/mile^2).[28] But density differences are pronounced within this mountainous landscape, where tillable soil is highly localized. Settlement patterns vary from dispersed to tightly nucleated, depending in part on the customs of various ethnic groups. Farmhouses and farming villages are often perched on hillsides or along ridges. Terracing is common, especially where land is scarce. Where flat or irrigated land is available, communities tend to locate on the edge of, but not on, the alluvium.

[26]Just to the south, the Puebla Project (as distinct from the Proyecto Puebla-Tlaxcala) did attempt to incorporate traditional agricultural practices.

[27]Samuel B. Bonis, *Geología del Area de Quezaltenango, República de Guatemala.*

[28]Dirección General de Estadística (Guatemala), *VIII Censo de Población y III de Habitación.*

Precipitation is, if anything, even more seasonal than in central Mexico, with more than 80 percent of the annual rainfall coming during the four months from June to September. Data from Quezaltenango (Xelajuj),[29] the region's principal city, suggest precipitation on the order of 900 mm/yr and average monthly temperatures from 12°C to 17°C.[30] But precipitation and temperature at any particular place are controlled strongly by topography, a situation requiring a dense network of local meteorological stations. Unfortunately, local stations are few, and local records are often broken and of doubtful accuracy. Thus, the climatic details of most places remain unknown.

For a general survey of soils in the region, I rely on the 1959 reconnaissance of Guatemala by Simmons, Tárano T., and Pinto Z.[31] Around the peaks of active and dormant volcanos, ashes and sands predominate (Map 3-2). Down slope but still above 2,300 m, relatively deep, well-drained sandy loams such as the Camanchá series have developed on volcanic ash on slopes of 5 to 12 percent gradient, and even up to more than 20 percent in some cases. Ostuncalco-series sandy loams have formed on newer ash, some of it from eruptions in this century.

Other ash-derived loams are found farther out and at slightly lower elevations. The easily eroded Sinaché-series clayey loams, for example, occupy some 7,000 ha on slopes of 5 to 12 percent gradient. Quezaltenango-series sandy loams are found on plains of less than 2 percent gradient at elevations of 2,250 m.

Narrow flood plains along streams may contain deposits of young alluvials. Although of minor extent, these undifferentiated mixtures of local and transported soils are often highly fertile and intensively managed.

Like their Tetlatlauhcan counterparts, the farmers of the *municipio* of Almolonga in the highlands of western Quezaltenango work both flat and sloping plots. From a narrow, nearly flat valley floor at about 2,250 m elevation, mountains rise abruptly and precipitously (Figure 3-2). Slopes are farmed to about 2,400 m, beyond which mixed oak/pine

[29]Something of a movement is underway to replace the Nahuatl Quezaltenango with the ancient Quiché name Xelajuj (also Xelaju, and in common usage the even shorter Xela). Jorge Luís Arriola (*El Libro de las Geonimias de Guatemala,* pp. 668–70) argues that the Mexicanization of Guatemalan place names took place well before the Spanish arrived, possibly as early as the tenth century.

[30]Observatorio Nacional (Guatemala), *Datos Meteorológicos de las Cabeceras Departamentales.*

[31]Charles S. Simmons, José Manuel Tárano T., and José Humberto Pinto Z., *Classificación de Reconocimiento de los Suelos de la República de Guatemala.*

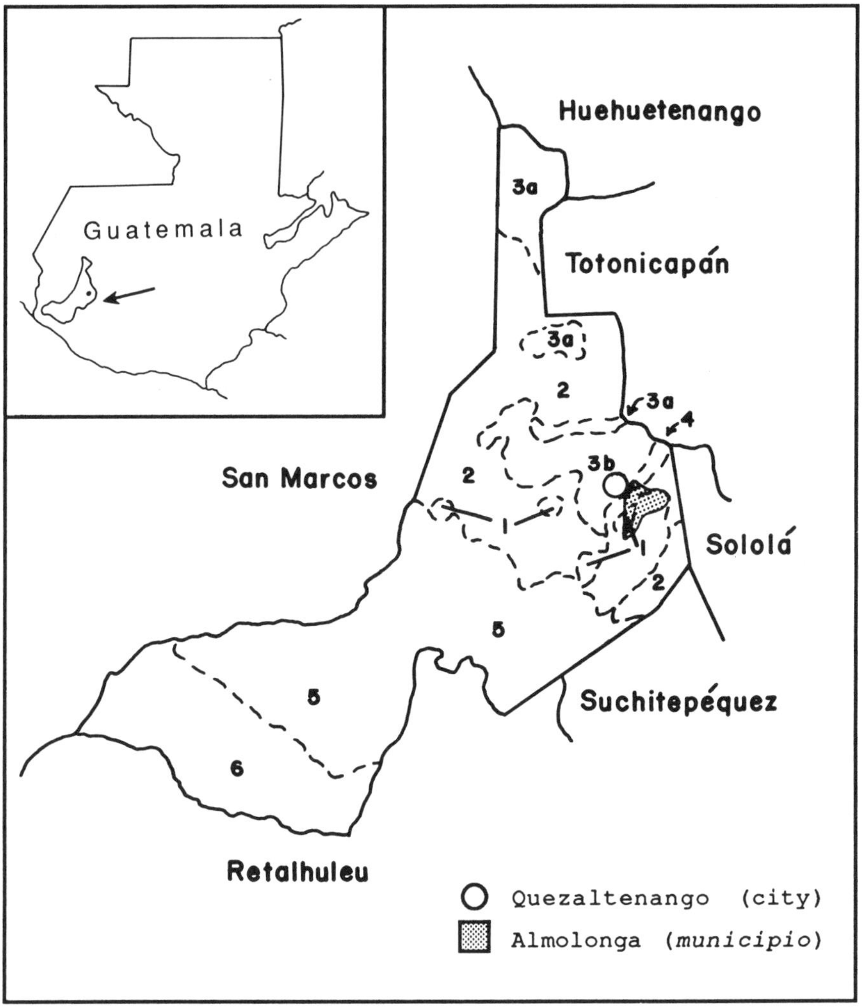

Map 3-2. Soils of Quezaltenango.

SOURCE: After Simmons, Tárano T., and Pinto Z., *Clasificación de Reconocimiento de los Suelos.*

Highland soils

1. Sands and ashes of volcanic peaks.
2. Sandy loams of volcanic mountains (Camanchá, Ostuncalco, and Totonicapán series).
3. Sandy and clayey loams of the *Altiplano*.
   - 3a. Soils on steep relief (Patzité, Quiché, and Sinaché series).
   - 3b. Soils on plains (Quezaltenango series).
4. Undifferentiated alluvials.

Other soils

5. Soils of the Pacific slope.
6. Soils of the Pacific littoral.

Figure 3-2. Landscape in Almolonga Valley, Guatemala. Intensive cultivation on valley floor and subsistence plots on surrounding mountain slopes.

forests predominate. Some peaks, such as nearby Cerro Quemado (2,818 m), are covered with bare lavas.

The *municipio* has a total area of 20 km² (2,000 ha), of which only 293 ha, or less than 15 percent, are classed as productive.[32] The remainder is mostly forested mountain land. Farms in this region are exceptionally small. The productive 293 ha are held in 635 individual farms *(fincas),* which average less than 0.5 ha each. More than 80 percent of the farms are less than 1 *manzana* (0.7 ha) in size and these average less than 0.3 ha each (Table 3-3). But even these figures do not fully convey a sense of the scale of farming in Almolonga. In the narrow valley where most of the people live, vegetables are grown on intensively irrigated and intensively managed parcels of *cuerda* (about 440 m²) size or less, and holdings of valley-floor land average only 0.15 ha (3 to 4 *cuerdas*).[33]

[32]Dirección General de Estadística (Guatemala), *II Censo Agropequario 1964,* vol. 2; Julio Cesar de la Roca, *Biografía de un Pueblo.*

[33]Dirección General de Estadística (Guatemala), *II Censo Agropequario 1964,* vol. 1.

Table 3-3. Use of Productive Lands, Almolonga

	Area (ha)	Farms (no.)	Average Size (ha)	Parcels (no.)	Average Size (ha)
Total productive lands	293	635	0.46	810	0.36
Less than 0.7 ha	141	516	0.27	627	0.22
0.7–1.4 ha	87	93	0.94	142	0.61
1.4–3.5 ha	38	19	2.00	29	1.31
3.5–7.0 ha	27	7	3.86	12	2.25

SOURCE: Dirección General de Estadística (Guatemala), *II Censo Agropequario 1964*, pp. 382, 419.

In 1973 the total population of Almolonga was 7,242, of which a surprising 6,281 people were listed as "urban." However, the one town, San Pedro Almolonga, in fact is a village, and most of its residents are farmers. Although gross population density of 362/km^2 (7,242 ÷ 20 km^2) (or 941/mile^2) is not spectacularly high by Middle American standards, population density in relation to farm land is extraordinary; there is only 0.04 ha per person in the *municipio*.[34]

Dense agricultural populations and intensive resource management are well correlated. Thus, it is not surprising that the Quiché farmers of Almolonga have developed a detailed soil classification and evaluation system. Almolongan farmers use two color and two textural classes to identify soils: *amarilla* (yellow), *negra* (black), *arena* (sand), and *barro* (clay). A third group, *tierras coloradas* (red soils) are not common in Almolonga but are common in the adjacent *municipio* of Quezaltenango and often enter into farmers' discussions as soils with which they are familiar. Yet even with only two basic colors and textures, together with a few modifiers that are applicable to both (little, some, much), the possibilities for fine distinctions are quite great. In practice only a few of the many possibilities are needed or used.

Like the Tlaxcalans, Almolonga farmers distinguish soils largely by color and texture but make sharper distinctions between hill and valley soils. Hill soils are mostly sands and sandy loams derived from volcanic deposits and are differentiated by color. Generally they are planted to subsistence maize and receive little attention other than plowing and

[34]Dirección General de Estadística (Guatemala), *VIII Censo de Población;* Escuela Facultativa, *Tenencia de la Tierra,* pp. 43–51, 171–208. Population densities and farm sizes for the entire country are reviewed by Alfredo Guerra Borges, *Geografía Económica de Guatemala,* vol. 1, chap. 8.

cultivating and occasionally light applications of chemical fertilizers. Valley soils, however, are so alluvially mixed and modified by people that they must be distinguished by finely graded descriptions of color and texture. Valley soils are among the most intensively cropped in Middle America. A regime of continuous multiple cropping imposes demands that can be met only with massive applications of organic and inorganic amendments. Responses to treatments thus form a larger part of soil evaluations than they do in less intensive systems (chemical and physical data are presented in Table 3-4.)

Thus, there are two classification and evaluation systems in use in Almolonga. In the hills, relatively unmanaged soils are classified by color and texture and evaluated on the basis of inherent qualities. Valley-floor soils, however, are so mixed and altered that division into classes would be difficult. Instead, soils are described with combinations of modified texture and color terms (for example, *más barro y negro*) and are evaluated as much on their ability to respond to treatment as they are on inherent qualities.

Hill Soils

Tierra amarilla (yellow soil) is a sandy to sandy clay loam of moderate quality. It is probably the most extensive of the hill soils used for the production of subsistence maize. Farmers may apply small amounts of chemical fertilizer to increase yields not only of maize grain but also of stalks and leaves, which are valued for animal fodder. Manure from stabled cows and other animals is usually not spread on hill plots but is reserved for valley-floor fields.

Tierra arenosa (sandy soil) is a sandy loam of indifferent quality. Apparently the *tierras arenosas* are similar to the *tierras amarillas* except that they contain higher percentages of sand, probably deposited from the nearby volcano, Cerro Quemado.

Tierra negra (black soil) is a clay loam with textural characteristics similar to those of *tierra amarilla* but with higher organic content. Despite this composition, *tierras negras* are considered of only ordinary quality, perhaps because of the local farmers' concentration on valley-floor soils and on management rather than on the intrinsic qualities of the various hill soils.

Almolongans are masters of soil and water management in their valley-floor fields. Yet they do not terrace hill fields even though slopes range from moderate to steep. Fortunately the hill soils do not appear especially erodible, possibly because generally high porosities limit surface runoff.

Table 3-4. Properties of Farmer-Identified Soils, Almolonga

	pH	Cond.	O.M. (%)	NO_3-N (ppm)	P (ppm)	K (ppm)	Zn (ppm)	Fe (ppm)	Lime (%)	Sand (%)	Silt (%)	Clay (%)	Texture	meg/ 100 g
Hill soils														
Tierra amarilla	6.5	0.2	2.8	1	10	283	1.5	+40.0	Low	48	27	25	Sandy clay loam	24.4
Tierra arenosa	6.7	0.2	1.3	2	21	134	1.4	+40.0	Low	73	15	12	Sandy loam	11.8
Tierra negra	6.4	0.1	4.0	1	22	175	0.8	21.4	Low	43	27	30	Clay loam	27.0
Valley-floor soils														
Untreated *media negra* soils	8.4	0.7	1.9	3	21	148	1.6	+40.0	Low	78	12	10	Sandy loam	12.0
Treated *media negra* soils	6.9	0.4	8.0	27	+80	325	+10.0	+40.0	Low	72	15	13	Sandy loam	28.8
Algo arenosa	6.5	0.4	10.8	70	+80	405	+10.0	+40.0	Low	72	16	12	Sandy loam	36.8
Más barro y negro	7.1	0.4	7.2	20	66	80	6.3	+40.0	Low	68	18	14	Sandy loam	28.2
Pura arena	6.9	0.3	4.5	51	+80	95	+10.0	+40.0	Low	72	16	12	Sandy loam	19.0
Pura amarilla	6.0	0.5	3.8	99+	52	133	4.5	+40.0	Low	65	21	14	Sandy loam	21.6
Puro barro o negro	6.2	0.5	9.6	33	+80	70	+10.0	+40.0	Low	60	25	15	Sandy loam	37.4
Description varies	7.2	0.4	4.3	9	57	405	4.6	+40.0	Low	58	27	15	Sandy loam	30.4

Valley-Floor Soils

The artificiality of valley-floor soils complicates analysis and discussion. Basically the soils consist of the same volcanic materials found in the hills but modified by alluvial deposition to form complex mixtures. The real differences, however, are due to human alteration. Only intensive production of high-return crops is justified on the scarce and enormously expensive land.[35] Products from the valley are sold locally in Quezaltenango markets and are also distributed widely in western and central Guatemala and are even exported to El Salvador and Mexico. From an average farm on the valley floor of 0.15 ha, a farmer produces enough to cover family expenditures (not including costs of food staples such as maize and beans, which are grown on hill plots), build savings and investments, and purchase the chemical fertilizers, pesticides, and often organic amendments, too, that are required to sustain continuous production. Organic amendments include manure from stabled and house-yard animals and leaf litter *(broza, hojarasca)* from nearby hills. Leaf litter is particularly valued as field dressing either by itself or mixed with animal manure and urine. The purposes and results of massive applications are highly altered soils.

Because few places on the valley floor are not subjected to such treatments, it is difficult to get samples or farmers' evaluations of "natural" soils. One opportunity was presented by a permanent raised field border along the edge of a small plot *(tablón)*.[36] The border had escaped most of the chemical and organic amendments, whereas the plot received heavy applications of both. Samples taken from the border (D-12) and the plot (D-11) show striking differences. Although both soils are blackish *(media negra)* sandy loams of basically similar textural characteristics, the effects of intensive soil management are readily apparent. Levels of nitrogen and of most other fertilizer minerals are much higher in the plot than in the border. The content of organic materials in the plot soil is exceptionally high, a characteristic shared with several other valley-floor soils that are periodically treated with *broza* and manure.

Because of alluvial mixing and human modification, valley-floor soils do not fall into easily identifiable classes. Colors and textures are

[35]Several informants at different times stated that valley-floor plots in Almolonga were worth Q500 to Q2000 per *cuerda* (Q1.00 = US$1.00).

[36]The term *tablón* (literally, plank or board; by analogy, any flattened planting bed) is applied to a wide variety of field forms in Guatemala, from the narrow unirrigated wheat terraces of Totonicapán to the intensely worked irrigated vegetable plots found on the deltas, hillsides, and valley floors of Sololá. If one adheres to local usage there is no escape from this confusing terminology.

used descriptively rather than as class names. Thus, the plot and border samples were described as *media negra* (blackish) without connoting a proper name. Perhaps Almolongan farmers, like soil scientists, acknowledge the undifferentiable nature of valley alluvials[37] and simply note fine differences between individual samples. To avoid the suggestion of type names, the valley soils are identified here only by sample number and description.

Six valley-floor soils besides those from the plot and border were sampled, rated by local farmers, and later analyzed. They appear in the following listing and in Table 3-4 in rank order from best to worst. My previous comparison of treated (plot) and untreated (border) soils will serve as a cautionary reminder; these are anthrosols, heavily modified by regular applications of chemical and organic amendments. Their characteristics are as much products of management as of natural endowments. Thus, a relatively poor soil may reflect less management for reasons other than inherent qualities, such as location in relation to irrigation water. All samples are from the surface (0 to 15 cm).

Sample D-22 (*algo arenosa,* somewhat sandy) is from a heavily managed plot. Regular applications of *broza,* manure, and chemical fertilizers are reflected in high levels of minerals and organic matter. Moisture absorption and retention characteristics are also of particular interest to farmers in this region of manual irrigation and year-round production. The present crop of cauliflower was planted immediately after a final harvest of potatoes.

For Sample D-23 (*más barro y negro,* more clay and black), the description does not refer to previous samples but to a cognitive norm for the valley. Thus, this sample is somewhat darker and has a higher clay content than the "average" valley-floor soil in the estimation of farmers. Although *broza* and manure were applied in past seasons, the present crop of potatoes was planted with only chemical fertilizers.

Sample D-24 (*pura arena,* pure sand; the designation is accepted hyperbole and indicates merely a high sand content) is texturally almost identical to Sample D-22, but the sandy nature of this soil dominates the farmers' description, possibly because levels of organic matter are lower. Such differences occur as a result of varying application rates of *broza* and other organic amendments, either because of economic constraints or because some farmers feel that less organic matter is adequate for particular crops, in this case successive plantings of lettuce.

Sample D-25 (*pura amarilla,* pure yellow) is a light-colored soil from the valley edges, slightly above the floor, on flat to gently sloping surfaces. Because it is away from surface-water supplies, it is not

[37]Simmons, Tárano T., and Pinto Z., *(Clasificación de Reconocimiento de los Suelos)* use the phrase *"suelos aluviales no diferenciados."*

irrigated or treated with *broza*. Occasionally fields in this area are dressed with manure, but usually only chemical fertilizers are applied to the customary maize crops. Thus, although this soil resembles such possible hill parent materials as *tierra amarilla*, it differs markedly in mineral content and texture.

Sample D-26 (*puro barro o negro*, pure clay or black) is an indifferent soil, but its location merits full management. Plot-side canals permit hand irrigation, and applications of *broza*, manure, and chemical fertilizers meliorate the soil's natural deficiencies so that it can be pressed into intensive use.

Sample D-27 (not identified with consistent descriptive terms) is a poorly regarded valley-floor soil that usually does not merit treatment with *broza* or manure. Respectable maize crops are secured with applications of chemical fertilizers, but plots are not planted to vegetables.

Summary

Because the soil-classification systems considered here are from regions of uncommonly intensive cultivation, they should not be considered typical of Middle America. Nevertheless, they well illustrate the levels of complexity achieved in some traditional societies, especially where farmers are skilled in resource management. With a few sensibly determined characteristics, such as color and texture, farmers can quickly establish major soil types. By combining characteristics and adding modifiers, they can describe a great range of variations. Because sensible soil characteristics are associated with properties of fertility, workability, and response to amendments, the same descriptive range also permits subtle evaluations of both the inherent and the managerial properties of soils.

A final classificatory step is identification of external factors, such as distance from the village, degree of slope, presence of rocks, and availability of irrigation water. Customarily these factors are considered more economic than edaphic, related to costs of cultivation and management rather than inherent soil qualities. But in both Tetlatlauhca and Almolonga less intensive cultivation and soil management on remote or hill lands are reflected in less detailed soil classification. Thus, classification emerges as a facet of traditional resource management, matching in complexity the cultural significance of the resource involved.[38]

[38]Brent Berlin, Dennis E. Breedlove, and Peter H. Raven, "Folk Taxonomies and Biological Classification."

4

Soil Management Using Organic Amendments

Soils are not necessarily depleted by the continuous growth of either wild or economic plants. But in natural stands nutrients are cycled from soil to plants and back, whereas in cultivated fields organic matter and nutrients are diverted to homestead, stable, and market. Having thus interrupted vital links, farmers must perform ecosystem management and work by determining and replacing needed materials.

In industrialized farming systems, resources for soil management, such as fuels and chemicals, are drawn from great distances through the supply lines of economically integrated societies. Materials from local sources are of little concern. In contrast, low-energy societies lack facilities for transferring and exchanging materials,[1] and resources come mostly from the immediate region. Over the years, farmers have learned to recognize and utilize locally available resources, at times to a degree astonishing to outside observers.

But if traditional farmers are spatially and economically isolated, they are linked to farmers everywhere by the universality of their problems and opportunities. The physical and biological aspects of farming are transcultural; such factors as nutrient uptake and moisture stress know no social or political barriers. The difference between more- and less-advanced farming systems is largely in tools and materials and not in basic requirements. Traditional farmers, no less than their

[1]Gonzalo Aguirre Beltran, *Regiones de Refugio,* pp. 95–96.

technologically advanced counterparts, must meet certain plant needs or risk poor yields or crop failure.

Thus, we need not look for distinctive soil-management problems in traditional farming but rather for distinctive solutions. Traditional farmers' reliance on locally available information, energy, and materials might seem to limit alternatives. But local wild and domesticated ecosystems offer a surprising range of effective soil amendments. Unlike highly concentrated manufactured goods, however, local materials tend to be bulky. Transporting and incorporating them into the soil pose problems that call for special solutions.

Because many soil-management techniques are exempt from economies of scale and because most soil amendments are infinitely divisible, small farm size does not necessarily reduce the number of soil-management options. In fact, it may increase them. In the intimate environment of a 1- or 2-ha plot, certain practices can be implemented that would be impossible on large fields.

A long list of topics could be addressed in a discussion of traditional soil management. This chapter focuses on the use of manures, plant residues, and other organic amendments to improve the fertility or physical properties of soils rather than on the nature and properties of the soils themselves.[2]

Manures

Animal Manures

Manure is an ancient and widespread soil amendment, used wherever corralled or stabled animals are part of the farming system.[3] In the centuries since large domesticated animals were introduced to Middle America, traditional farmers have enthusiastically adopted the practice of manuring.[4] Uses of dung for other purposes, such as fuel or as binder for adobe blocks, are rare; most manure goes onto farm fields, and application is limited almost entirely by the relative scarcity of animals. Farmers who have animals are considered fortunate to have a ready supply of manure for dressing fields. Those who do not are

[2]For a general review of this subject and a number of specific studies, see Food and Agriculture Organization, *Organic Material as Fertilizers.*

[3]Eugene Mather and John Fraser Hart, "The Geography of Manure."

[4]The few domesticated animals in pre-Spanish Middle America did not congregate in such a way as to produce substantial accumulations of manure and neither do most wild animals. The deposits of those that do, such as bats and sea birds, were exploited.

restricted by the supply of manure for sale or by lack of funds for purchase.

All types of manure are used. Animals that spend much of their life in confinement, such as cows or pigs, are especially good sources because manure collection is simplified. But even wide-ranging animals are valued as manure sources. For example, folding sheep in rotation on fields is common in southern Mexico.[5]

Perhaps the most common source of manure for traditional farms is the house yard itself. Over much of Middle America, rural dwellings are not single buildings but collections of structures and open activity spaces. These range from unenclosed clusters of sleeping, cooking, and storage structures and stables to linked and in some cases attached buildings surrounded by fences or walls.

This tight clustering of human dwellings and animal shelters produces inevitable accumulations of waste: animal droppings, straw and other bedding material, remnant maize stalks and cobs, feathers from house-yard poultry, and whatever else finds its way into the compound and is not consumed. Periodically the manure is cleaned from beneath the animals and stored, usually outside the yard, in piles or pits. Kitchen refuse *(basura)* and yard sweepings are added.[6] Ash *(ceniza)* from kitchen fires is thrown on the piles or is set aside to be dusted over young maize or vegetable plants for treatment or prevention of fungus *(chahuistle)*.[7] Thus, manure piles become layered compost heaps in which organic materials partially decompose to produce high-quality field dressings.

Bedding materials are particularly important because they absorb liquid wastes, especially nitrogen-rich urine (Table 4-1), and control decomposition and nutrient loss in the storage piles. Traditional farmers in Middle America use a variety of materials, from carefully spread straw to casually developed fodder remnants that are left after leaves and other edible parts are consumed. Maize stalks and other residues are broken and mixed by the treading action of the animals themselves.

In intensive systems, bedding material and manure may be carefully formulated. Thus, for the carefully worked vegetable plots *(tablones)* of Quezaltenango, Guatemala, straw is saved or even purchased and placed beneath stabled cows for a week or so, then transferred to storage piles or to fields, where it is interred with broad hoes *(azadones)*. Even more

[5]Evon Z. Vogt, *Zinacantan*, pp. 37–38; Kenneth Weathers, "La Agricultura de los Tzotzil de Nobenchauc, Chiapas, México."

[6]I have seen farmers salvaging garbage from city dumps to spread on fields after they first discard glass, plastic, and other nondegradable materials.

[7]The term *chahuistle* (from the Nahuatl *chiahuiztli*) is applied rather generally to a variety of fungus and insect infestations.

Table 4-1. Composition of Fresh Animal Excrement

Excrement	H_2O (%)	N (%)	P_2O_5 (%)	K_2O (%)	Excreted per Year[a] (M tons)[b]	Weight of Bedding (M tons)[b]	Total Weight (M tons)[b]
Horse							
Solid, 80%	75	0.55	0.30	0.40			
Urine, 20%	90	1.35	Trace	1.25			
Whole manure, 100%	78	0.70	0.25	0.55	9.00	3.0	12.00
Cow							
Solid, 70%	85	0.40	0.20	0.10			
Urine, 30%	92	1.00	Trace	1.35			
Whole manure, 100%	86	0.60	0.15	0.45	13.50	1.5	15.00
Sheep							
Solid, 67%	60	0.75	0.50	0.45			
Urine, 33%	85	1.35	0.05	2.10			
Whole manure, 100%	68	0.95	0.35	1.00	6.25	3.5	9.75
Pig							
Solid, 60%	80	0.55	0.50	0.40			
Urine, 40%	97	0.40	0.10	0.45			
Whole manure, 100%	87	0.50	0.35	0.40	15.25	3.0	18.25
Hen							
Whole manure, 100%	55	1.00	0.80	0.40	4.75	—	4.75

SOURCE: After Lucius L. Van Slyke, *Fertilizers and Crops,* Chap. 17. See also A. L. Sutton et al., "Utilization of Animal Waste as Fertilizer."

[a]Fresh manure without bedding.
[b]Tons/500-kg live weight of various animals.

distinctive of this region is the practice of placing leaf litter gathered from nearby mixed oak/pine forests beneath stabled animals. After a week or so the rich mix of pulverized leaves, manure, and urine is spread over the tiny plots and turned under.

Various types of manure have qualitative differences in the opinion of traditional farmers. Sheep and goat manures are rated highly because they are "rich" *(rico)* and tend not to "burn" *(quemar)* the crop plants (that is, produce large amounts of ammonium carbonate and its nitric-acid compounds). Cow manure is the most abundant. Only rarely is pig, burro, mule, or horse manure available in large concentrations. Usually

Figure 4-1. Scattered piles of manure ready for spreading. Application will be light even though sandy soils would benefit from heavier dressing.

waste from these animals becomes a minor component of the general farmstead manure pile, along with droppings from chickens, turkeys, and other small animals. Similarly, farmers may express strong preferences for different kinds of manure for specific crops. Thus, herb and spice growers in Oaxaca generously dress their plots with goat manure even while complaining of high prices.

Availability determines manure application rates. Fortunate farmers with several cows, horses, burros, or other stabled animals may have enough to satisfy the requirements of their small fields. Those with fewer animals apply what they have (Figures 4-1 and 4-2).[8] The problem has two aspects: supply and distribution. In many communities manure is scarce and jealously guarded by those with producing animals. But even if surpluses existed nearby, transportation costs and difficulties would rule out extensive transfers. A few of the more affluent traditional

[8]Few traditional farmsteads have enough animals to produce sufficient manure for even hectare-sized fields. Instead, available manure is concentrated on small, intensively cultivated plots, while larger, more remote fields go unfertilized. Despite the sometimes high application rates, the small plots are probably not overmanured because the common practices of multiple cropping and intercropping place heavy drains on soil nutrients.

Figure 4-2. Heavy dressing of manure by farmer with relatively large herd of cows.

farmers own or hire trucks. But most must rely on animal-drawn carts or wagons or even panniers for cartage (Figure 4-3). These slow conveyances with their limited capacity are inadequate for transporting the sizable volumes of material involved in manuring fields.[9]

There are exceptions to these constraints. Many traditional farmers have partially commercialized operations and grow specialty vegetable, herb, or beverage crops for urban markets. These crops may require fertilizer to be commercially acceptable, but the farmers enjoy a relatively high cash income with which to purchase the necessary materials. For example, farmers in the fertile Tepeaca and northern Puebla valleys of Mexico have long grown vegetables, herbs, and flowers for local markets and for shipment to Mexico City. They prefer chicken manure, especially on light, sandy soils, where chemical fertilizers are used sparingly. A good supply is available from the big egg and dressed-poultry ranches *(granjas)* of Tehuacán. But Tehuacán is 90 km distant from the vegetable and flower districts of Tepeaca and more than 150 km from the vegetable and condiment fields near Puebla and in

[9]Manure applications decline sharply on fields more than an hour from sources, which is not a great distance by animal-drawn wagon. Michael Chisholm, *Rural Settlement and Land Use*, pp. 31, 53ff.

Figure 4-3. Bottom-dumping, burro-mounted panniers used for transporting manure to fields.

southern Tlaxcala. Nevertheless, farmers in these regions are willing to pay the equivalent of US$8 to $12/ton for truckloads of chicken manure, which they feel is essential for their high-quality, commercial crops. Similarly, vegetable growers in Quezaltenango order chicken manure trucked more than 100 km from the *granjas* of Guatemala City.

Human Waste

Although the use of human waste had a long and honorable history in pre-Spanish Middle America,[10] it is now a subject to be broached with some delicacy. Farmers universally and vehemently deny that night

[10]Human waste was systematically collected in the Aztec capital of Tenochtitlán and transferred by boat to tanneries and produce gardens. Pedro Armillas, "Land Use in Pre-Columbian America"; Brigitte B. de Lameiras, *Terminología Agrohidráulica Prehispánica Nahua,* pp. 38–40; Bernal Diaz, *The Conquest of Mexico,* p. 233. In a modern version of this ancient practice, treated sewage is pumped from plants in Mexico City to city parks and to farmlands near Tula, Hidalgo. Richard H. Sullivan, Morris M. Cohn, and Samuel S. Baxter, *Survey of Facilities Using Land Application of Wastewater,* pp. 138–40.

Table 4-2. Composition of Ant-Nest Refuse (Oaxaca)

Sample	N (%)	P (%)	K (%)
D-50c	2.16	0.41	2.28
D-53	2.41	0.18	0.09

soil is used in agriculture. Nevertheless, technological and spatial factors ensure that much of the human waste generated in farming regions is returned to the fields.

Central plumbing is rare in traditional farmhouses and villages. Often even latrines are lacking. Under such circumstances, the most common solution for villagers is to seek out the nearest maize field, canal bank, or other semiprivate location. Around farmhouses, the adjacent maize field or dooryard garden serves the family. In some instances the manure pile itself offers a convenient depository for human waste. In the *chinampa* district of Mexico, and in other areas laced with open canals or drains, the waterways receive deposits either directly or as runoff from fields and villages. Some of this material eventually returns to the fields when they are irrigated or mucked with canal bottom deposits.

Ant-Nest Refuse

A final manure deserves mention, even though it is used in only a few places and on a limited scale, because it illustrates the minute recognition of local resources that typifies traditional farming. In parts of the Oaxaca Valley of Mexico, debris ejected from the nests of gardening (also leaf-cutting) ants (*Atta* spp.) is highly prized by farmers for fertilizer. The nutrient-rich material (Table 4-2), which resembles fine sawdust, accumulates in substantial mounds around tree trunks or ground nests. It is the depleted organic substrate of the fungus gardens that support attine colonies and consists of partially rotted plant material plus fungal and fecal matter.[11]

Use of ant refuse is highly localized. Near the Oaxaca Valley town of Zaachila, the material is assiduously collected, to the point of contention when one farmer trespasses on another's land in search of nests. The material, here called *caca de arriera* (*caca,* a child's term for excrement, and *arriera,* driver of freight horses or mules; thus, one who

[11]Caryl P. Haskins, *Of Ants and Men,* pp. 39ff.; Neal A. Weber, *Gardening Ants: The Attines,* pp. 6-7, 93-94.

drives or hurries, a reference to *Atta's* propensity to run), is gathered in bags, then carefully measured out one or two handfuls at a time and spread around individual plants or plant groups *(matas)* of high-value crops such as green (husk) and red tomatoes, chile, and onions. In the Zapotecan village of Yalalag, more than 60 km northeast of Zaachila in the rugged Mesa del Sur, *excremento de la hormiga* (ant) *arriera* is used specifically on chile seed beds and planting beds.[12]

Elsewhere ant refuse is known as a fertilizer but is little used. Thus, around Ocotlán, less than 20 km from Zaachila, substantial refuse mounds accumulate beneath *guamúchil* (*Pithecellobium* spp.) trees but are ignored by passing farmers. Although they agree that the *caca*—here also called *excremento* or *suciedad* (filth) *de arriera*—is a rich soil amendment, they argue that it is not worth the trouble of collecting.

The ant is also well known in Huehuetenango and Quiché, Guatemala, where its deposits are called *abono de zampopo* (manure of zampopo ants), indicating local familiarity with its fertilizing properties. However, it is collected and applied only occasionally to maize, chile, and onions. In this arid region, the zampopo also harvests leaves of crop plants, so colonies are viewed with mixed feelings as both resources and hazards.[13]

The same situation applies in other areas. Farmers in Quezaltenango and Sololá, Guatemala, recognize ant refuse as fertilizer but claim there are few or no colonies in their immediate area. Far to the north, in the Tehuacán Valley of Mexico, refuse mounds develop beneath *nopal (Opuntia)* cactus and *mesquite* (*Prosopis* spp.) trees. Farmers call it *abono de los animalitas* (manure of the little animals) and confirm that it is good fertilizer but add that there is so little that no one uses it.

Widespread recognition of ant refuse as manure suggests that the practice may have been more common in times past. Present limited use is probably a result of substituting commercial fertilizers for ant refuse,[14] especially in those areas where wage rates and income levels are relatively high. Supplies of *caca de arriera* never could have been adequate for general farming use, such as for field maize, and it seems likely that the near-universal adoption of chemical fertilizers by farmers who cultivate intensively has diminished interest in this curious organic soil amendment.

[12]Julio de la Fuente, *Yalalag*, p. 85.

[13]Ants may also have a role as biological control agents. Margaret E. Saks and C. Ronald Carroll, "Ant Foraging Activity in Tropical Agro-Ecosystems."

[14]One farmer near Aguacatán, Huehuetenango, said that chemical fertilizers sometimes were mixed with *abono de zampopo* before application.

Plants

Plant Residues

Probably the most common soil amendments in Middle America or, for that matter, anywhere in the traditional farming world are the whole, chopped, or burned remains of previous crops, weeds, or fallow growth. Unfortunately, demands for fuel or fodder have driven many to clean harvesting. But where animals are scarce or alternative fuel and fodder sources are abundant, the ancient practice of using ashes and plant refuse as soil conditioners continues.

Ash

The use of ashes is prevalent in the most animal-scarce, fuel-abundant traditional system, shifting cultivation.[15] Although there are many variations, the practice essentially consists of cultivating scattered holdings in areas of natural vegetation and then abandoning them as yields decline. Fallow periods, during which the vegetation and soils recover, exceed cultivation periods, usually by a factor of five or more. Shifting cultivation is also called forest or bush fallow to suggest both the nature of the natural vegetation and the characteristic of field rather than crop rotation.[16] Competition from other forms of agricultural land use has relegated shifting cultivation to the rough lands of Middle America. Fields are often broken, rocky, and strewn with debris after burning, which makes them unsuitable for draft animals and plows.

Site selection is based in part on direct observation or use of plant indicators to determine soil capabilities. After cutting, abundant vegetation is left to dry where it falls. When vegetation is uneven or sparse, the slash may be spread to cover the ground evenly to secure a good burn or piled to concentrate the fuel.[17]

When the cut vegetation is burned, most of the nitrogen and sulphur are volatilized, while other nutrients remain on the soil surface in the ash.[18] The ultimate amount of nutrients delivered to the crops depends on several variables, such as volume and composition of the

[15]Shifting cultivation is also called slash-and-burn, swidden (a transfer from Europe), *roza,* and *milpa,* although *milpa* actually refers to maize in the field or the maize field itself. E. Eckwall, "'Slash-and-Burn' Cultivation"; O. F. Cook, "Milpa Agriculture."

[16]Emil B. Haney, Jr., *The Nature of Shifting Cultivation in Latin America;* R. F. Watters, *Shifting Cultivation in Latin America.*

[17]William E. Carter, *New Lands and Old Traditions,* pp. 20–37, 43–46; Haney, *The Nature of Shifting Cultivation;* Watters, *Shifting Cultivation.*

[18]P. H. Nye and D. J. Greenland, *The Soil Under Shifting Cultivation,* pp. 66–70.

Figure 4-4. Wheat stubble and burned terraces, Totonicapán, Guatemala.

original stand, completeness of burn, and weather. Because shifting cultivators do not inter the ash, wind or sheet wash can carry it off before the nutrients have leached into the root zone. Blowing or washing away occurs especially with the lighter "white ash" produced by very hot fires, as compared with "black ash," which more readily enters the soil.[19] In addition to producing a fertilizing ash, burning may also destroy seeds and weed sprouts, ants and other insects, microbial populations, and organic matter in the soil, and may generally upset microhabitats and displace their occupants.[20]

Despite its extensively documented properties, the fertilizing ash from slash-and-burn farming is more a by-product of clearing than a deliberately created soil amendment. The critical transfer of ash from surface to root zone is left up to leaching by fortuitous rains. In some cases of poor burns (for example, slash soaked by premature rains), farmers have complained to me more of field working conditions and weed growth than of nutrient losses. In fact, the basic problem faced by

[19]Carter, *New Lands,* p. 55.

[20]Secondary effects of burning on the soil are related to temperatures developed below the surface, which in turn depend on temperature and duration of the burn and insulating characteristics of the soil. Nye and Greenland, *The Soil Under Shifting Cultivation,* pp. 70–73.

shifting cultivators is climatic, not edaphic; they must destroy the shading forest to create the high-energy surfaces needed by their sun-loving crops. In low-energy systems, the most expedient way to destroy the forest is by cutting and burning. The resulting ash is thus an adventitious benefit of microclimate modification.[21]

But shifting cultivators are not the only ones to use ashes as soil amendments. Burning drastically reduces the bulk of plant residues and thus the depth to which soils must be turned. It is a common practice, especially where remains from dense crops such as wheat or oats are adequate for continuous combustion. When small grains are harvested, often by hand, straw is left about 50 cm high to provide fuel for subsequent burning (Figure 4-4). In cases where dry residues are scarce, burning may still be practiced. To develop a sufficient concentration of fuel, crop residues and weeds are raked into piles and fired; the resulting ash is redistributed over planting surfaces. Even more deliberate than this procedure is the transfer of ash from domestic cooking and heating fires directly to the fields or indirectly via house-yard trash heaps. Domestic ash is often carefully applied by hand around each plant, suggesting the high regard farmers have for this amendment.

Plant Remains

The widespread practice of interring plant remains varies from the rather casual turning under of stubble and weeds in the normal course of field preparation to the careful collecting and placing of organic materials in fields to ensure maximum effectiveness. But burying crop debris and other materials involves turning soils deeply enough to completely cover the residues and to leave a clean surface for planting. Burying debris this deeply poses a problem for low-energy, traditional systems, which use hoes or animal-drawn plows to work soils.

A number of solutions are found in Middle America. Burning to reduce bulk has already been discussed. Alternatively, modifying field surfaces facilitates interment of plant remains. Traditional fields are often formed into ridges, furrows, raised planting beds, or other forms during plowing or subsequent cultivation. For example, maize and associated crops such as beans are commonly first planted in plow-formed furrows that are later reformed into ridges. Stubble and other plant residues that remain on the ridges after harvest are turned over into the furrows during field preparation for the next cropping period. The ridge-and-furrow configuration facilitates interring plant remains with relatively shallow-cutting animal- or hand-powered tools. Where

[21]Kekchi cultivators seek five principal results from burning: fields cleared of slash, loosened soils, lowered rates of weed regrowth, reduced animal and insect pests, and, only lastly, nutrients for crops. Carter, *New Lands*, pp. 54–55.

Figure 4-5. Maize stalks laid in furrows between mounds prior to chopping and turning under.

Figure 4-6. Method of burying stubble in hoe-tilled fields, Totonicapán and Quezaltenango, Guatemala.

fuel and fodder demands for crop residues are not too heavy, even maize stalks and other large plant remains are left in the fields. In such cases it is common to place the debris in old furrows, chop it with *machetes* (long knives) or hoes, then inter it along with the stubble as described (Figure 4-5).

Small grains such as wheat or oats are usually planted on flat, rather than ridged or mounded, fields. Even if burned, the stubble is difficult to bury, especially for farmers who lack draft animals. Hoe farmers in Quezaltenango, Guatemala, attack this problem by first opening narrow furrows about 30 cm deep, into which successive cuts of root-matted soil

Figure 4-7. Turning under wheat stubble with the *azadón*.

and stubble fall. The plowing (*surcando,* from *surcar,* to furrow or plow) is accomplished with two swings of the *azadón;* the first swing cuts the sod and drops stubble and root mats into the furrow, the second cut lowers the furrow and pulls fresh dirt over the debris (Figures 4-6 and 4-7). The farmers thus are creating ephemeral ridges and furrows and working with gravity to bury old planting surfaces.

On narrow hand-worked terraces *(tablones),* wheat farmers of Totonicapán, Guatemala, employ an ingenious system for incorporating stubble into new planting beds. Rather than disturbing the previous season's surface, farmers shave soil from terrace risers with *azadones* to cover old surfaces and stubble or ash (Figure 4-8). By burying old surfaces in this manner, farmers inter organic materials at root levels of the new crop and reduce the potential for erosion because basic terrace structures remain undisturbed and reinforced with intact root systems.

In the several methods described for dropping soil and plant remains from old ridges or plot surfaces into furrows, primary transfers are downward, assisted by gravity. On flat fields, the potential for using gravity is renewed when ridges are reformed. But they are reformed after plowing, a much easier task than when soils are compact. On sloping fields, the gravity debt is repaid only indirectly by the energy expended transferring the farmer and his tools up slope.

Figure 4-8. Burying old *tablón* surface, Totonicapán, Guatemala.

In most systems, weeds that have grown since the last cultivation are simply turned under along with whatever crop residues remain on the surface. Especially noxious weeds, however, such as those that sprout from rhizomes or stolons, are separated out for disposal. On some intensively cultivated plots, nonnoxious weeds extracted by hand or hoe during cultivations are set aside in piles, then returned as mulch at next plowing.

Green Manures

Cover cropping is not widespread in traditional Middle America. When climate permits a crop, farmers usually plant something that can be eaten and rely on crop residues, manures, or plant mulches to add organic materials to soils. Exceptions occur with soils particularly deficient in organic matter. For example, when converting tuffacious *tepetate* to soil, farmers may first plant a cover crop that will be turned under before regular cropping on these new soils begins. However, such crops are usually planted only once, during the initial conversion process.

Figure 4-9. *Sauco* stumps dotting sandy fields near Ostuncalco, Guatemala.

A distinctive and considerably more systematic form of green manuring than cover cropping is practiced near Ostuncalco, Guatemala, where farmers maintain *sauco (Sambucus mexicana)* trees in their potato, maize, and bean fields. The *sauco* are not allowed to achieve tree stature but instead are pruned rigorously so that only small stumps are left to sprout new growth (Figure 4-9). Leaves and small branches are removed annually, spread carefully around individual plants or *matas*, then chopped and interred with broad hoes. Farmers claim that crop quality and yields depend on these annual applications of *sauco* leaves. Soils in the Ostuncalco region are derived from volcanic ash; they are sandy and highly erodible, with rapid internal drainage and low moisture-retention capacity.[22] Generally they are low in nitrogen and organic materials. *Sauco* leaves are relatively high in nitrogen (Table 4-3) and decompose readily when fresh. With a low carbon/nitrogen ratio, they should promote rapid organic turnover in the soil and yield their nitrogen quickly.[23]

[22]Charles S. Simmons, José Manuel Tárano T., and José Humberto Pinto Z., *Classificación de Reconocimiento de los Suelos de la República de Guatemala*, pp. 167–69, 847.

[23]Nyle C. Brady, *The Nature and Properties of Soils*, pp. 434ff.

Table 4-3. Composition of Leaves

Type	N (%)	P (%)	K (%)	Ca (%)	Mg (%)
Sauco (*Sambucus mexicana*) leaves[a]	2.64	0.77	1.50	—	—
Broza (leaf litter)[b]					
Pino (pine)	0.71	0.09	0.10	2.23	0.30
Cipresa (cypress?)	0.83	0.11	0.10	1.98	0.42
Encino (oak) (test 1)	1.18	0.15	0.23	1.19	0.30
Encino (oak) (test 2)	1.05	0.06	0.26	0.58	0.52
Aliso (alder)	2.35	0.10	0.09	1.69	0.50

[a]Completely dried before analysis.
[b]Analysis by perchloric digestion.

Organic Mulches

If crop residues are insufficient or inadequate, additional materials may be taken from other fields or from wild habitats and applied as organic mulches. Mulches include fresh and dried plants and composts and are distinguished here from crop residues primarily by the fact that they are brought in rather than developed in place. Within the range of plants available in a region, farmers have some choice over which types to apply. But low value and difficulty of transport limit the distances over which bulky, organic matter can be carried, and the materials used for mulches reflect accessibility as much as preference for particular plants.

The variety of materials used for organic dressings is wide. Household refuse itself supplies a net gain in organic matter for those fields that receive more than they supply. Leaf litter from field-side trees or dooryard gardens is often collected and spread over fields. Crop residues from one field may be transferred to another. For example, in western Guatemala, wheat straw brought from thrashing floors is spread along old maize furrows prior to planting the new crop. In addition, mulches may be derived from materials used first to protect seed beds and delicate seedlings and then as soil amendments. Leaves, grasses, and branches originally used as arbors or surface mulches are often subsequently chopped or burned and incorporated into soils.

Mulches are used for various reasons. Some are of little nutritive value and primarily improve soil structure; others supply minerals that

Figure 4-10. *Chinampa* compost pile made from crop residues, aquatic plants, tree leaves, and weeds.

enhance soil fertility. But other factors also enter into the decision to mulch. For example, farmers in southern Tlaxcala, Mexico, spread *acacapa* (water hyacinth, *Eichornia crassipes*) over newly planted fields, then later bury it during cultivation. The procedure achieves three objectives. Drains *(zanjas)* vital to farming in this region of high water tables are cleared of clogging aquatic weeds, field surfaces are protected by mulches, and organic material is added to the soil.[24] Usually the hyacinth is left along canal or *zanja* banks to drain and wilt. Those that are manageable are then spread over fields, where they dry to a fraction of their original weight and bulk.[25] Thus, although the hyacinth initially forms a mat over plot surfaces, the final product is easily incorporated into the soil with team-drawn equipment.

A slightly different practice is followed in the *chinampa* district of Mexico, where aquatic plants scooped from canals are placed in heaps along with crop residues, tree leaves, and field weeds (Figure 4-10). The partially rotted material is later used to compost intensively cultivated

[24]Gene C. Wilken, "Drained-Field Agriculture."

[25]Some 95 percent of the fresh weight of *E. crassipes* is water. National Academy of Sciences, *Making Aquatic Weeds Useful*, p. 78.

Figure 4-11. Burying *broza,* Zunil, Quezaltenango.

vegetable and flower plots. Aquatic weeds, with their extraordinarily high water content, are ideal for composting.[26]

Mulching with introduced rather than local amendments can be considerably more selective and systematic. For example, in Quezaltenango, Guatemala, leaf litter *(broza, hojarasca)* is brought in large quantities from nearby forests to improve tilth and moisture retention of intensively worked vegetable plots. Some farmers own or control mountainous forest land from which they regularly harvest litter for their irrigated valley fields. Those who do not must purchase *broza* from truckers with access to forest lands.

In this procedure, litter is raked up from forest floors, placed in bags or nets, and carried to fields by men or horses or, from more distant sources, by trucks. After being spread over the fields, leaf litter is worked into the soil with broad hoes (Figure 4-11). In some cases the litter is first placed beneath stabled animals, then after a week or so the rich mix of pulverized leaves, manure, and urine is spread over plot surfaces and turned under. The urine tends to offset field losses of nitrogen caused by bacterial action in the decomposing litter.

[26]National Academy of Sciences, *Making Aquatic Weeds Useful,* chap. 9.

Use rates for leaf litter vary. Some farmers claim that they apply as much as 40 metric tons/ha/yr. Many use less. Perhaps a range of 20 to 30 tons/ha/yr would include most of the *broza* users.[27] Rough calculations made in mixed pine/oak stands near the city of Quezaltenango indicate that a hectare of forest produces about 4,000 kg of litter annually. Thus, in this region a hectare of cropped land requires the litter production from 5 to 8 ha of regularly harvested forest.

Traditional farmers who use *broza* are well aware of how to put it to best use. The highly regarded *broza* is not applied to the sandy soils of the hill slopes; it is reserved for the higher-quality alluvials of the valleys because even on these fertile soils the nearly continuous production of intercropped vegetables of commercial quality[28] is possible only with frequent applications of amendments.

Knowledgeable traditional farmers apply local amendments because they prefer them for certain purposes, not because they do not know of other alternatives. For example, Quiché (Maya) farmers of Guatemala are well informed about modern agricultural inputs and discuss at length the relative merits of various pesticides, purchased seeds, and chemical fertilizers in relation to specific crops and soils. But they are equally knowledgeable about local organic amendments. They know that sandy *(arenosa)* soils need manure whereas clayey *barros* benefit most from *broza*. Bedding mixtures of *broza* and manure can be used on any soil with good results. And even finer treatments are possible. Although most *broza* is a mixture as are the forests from which it comes, some leaves, such as oak *(encino)*, are preferred, and specific leaves are recommended for particular soils. For example, on easily worked *(suave)* soils, best results are obtained with alder *(aliso)* and cypress *(cipresa)* litter. On sticky or compact *(duro)* soil, pine *(pino)* needles and oak leaves are effective. Some clay soils contain lumps of *taxcal*, a cemented material probably of volcanic origin, which can be broken up with oak leaves.

Although not insignificant, the nutrients added by *broza* (Table 4-3) are of secondary importance. Farmers depend more on animal manures and chemical fertilizers to supply deficient minerals and maintain soil fertility under nearly continuous cultivation. The massive *broza* applications mainly improve soil structure and water retention and

[27]Data are presented in tons per hectare for convenience and comparisons. Actual plot sizes are extremely small, typically only 0.10 to 0.15 ha each. Thus each plot would receive only 10 to 15 percent of the indicated quantities of leaf litter.

[28]Data on vegetable production in Guatemala are contained in T. David Johnson, *Análisis de Actividades Necesarias para la Producción de 11 Especies Horticolas (Hortalizas).*

serve as carriers and buffers for animal wastes. Data developed elsewhere suggest the absorptive capacities of some common bedding materials and the excellent characteristics in this regard of leaf litter (Table 4-4).

Use of leaf litter is common throughout Middle America, although it is usually less systematic than in western Guatemala. Almost everywhere leaves from dooryard gardens or field-side trees are spread over field surfaces. As mentioned, in the *chinampa* district south of Mexico City leaves from abundant trees that line the narrow plots are combined in mulch piles with aquatic plants from adjacent canals and with crop-plant residues. Leaf litter occasionally is brought into the *chinampa* district, but I did not determine the source areas and extent of this practice.

Nitrogen Fixation with Plants

Although the effectiveness of planting crops that fix nitrogen is subject to controversy, the practice is thoroughly imbedded in traditional agriculture.[29] For example, intercropping beans with maize is almost a trademark of Middle American farming; the mix of crops is considered as essential to the vigor of the soil as the mix of foods is to the vigor of the people.[30]

But the practice goes well beyond slavish adherence to traditional subsistence patterns. Farmers have long recognized that certain plants improve soil conditions and increase yields of companion crops, and they plant accordingly. For example, farmers in Tlaxcala, Mexico, are convinced that tomatoes grown for market produce better if intercropped with common beans *(frijol, Phaseolus vulgaris)*. In Totonicapán, Guatemala, broad beans *(haba, Vicia faba)* or other legumes are scattered among the primary crop of wheat, also a commercial crop. In neither case does intercropping interfere with normal field operations because crops are cultivated and harvested by hand.

Similarly, alternating nitrogen-fixing legumes with field crops is common. Alfalfa, often grown in small plots to support a few stabled milk cows, is viewed as a valuable soil conditioner for maize or other grain crops. If land is particularly deficient (*debil* or *flojo,* weak, flaccid), a single crop of broad beans or other legume may be planted to "benefit the soil" *("se beneficen la tierra")*.

Less well known but still widespread is the practice of maintaining native leguminous trees in cultivated fields. In warm valleys from

[29]Brady, *Nature and Properties of Soils,* pp. 151ff.

[30]If eaten together, the traditional Middle American mix of maize and beans plus minor foods such as squash and chile provides all the essential amino acids. Nutritional catastrophe can result if the mix is not maintained. Food and Agriculture Organization, *Maize and Maize Diets.*

Table 4-4. Characteristics of Litter Materials

Material	Litter Required to Absorb 100 kg of Liquid (kg)	Fixing Capacity for Ammonia Nitrogen Held per M Tons of Litter (kg)	Content of Fertilizing Ingredients per M Tons of Air-Dry Matter			Content of Cellulose and Hemicellulose (%)
			Nitrogen (kg)	Phosphoric Acid (kg)	Potash (kg)	
Leaf litter	25–60	13.3 (oak)	8	3	3	28 (oak)
Wheat straw	45	2.25	5.5	2	10	65
Oat straw	35	3.55	6	2	13	High
Rye straw	45	1.7	6	3	8.5	62
Chopped straw	20–30	—	—	—	—	—
Maize stalks (shredded)	25–35	2.65	7.5	4	9	51
Sawdust	25	0	2	1	2	—
Wood shavings (softwood)	25	0	2	1	2	—
Peat moss (sphagnum)	10	20	8	1	1.5	22–37
Peat (sedge and woody)	15–25	15–30	10–30	1	1.5	2–10

SOURCE: Adapted from Thorne and Peterson, *Irrigated Soils*, p. 185.

Figure 4-12. Leguminous trees in a maize field, Tehuacán Valley, Mexico.

Puebla and Tehuacán south to Oaxaca, Mexico, farms with light to moderately dense stands of *mesquite* (*Prosopis* spp.), *guaje (Leucaena esculenta),* or *guamúchil* (*Pithecellobium* spp.) are familiar sights. Stand densities vary from a few scattered trees to veritable orchards, in which field crops appear as an understory (Figure 4-12). The trees are seldom deliberately planted. Instead, volunteer seedlings are protected during plowing and cultivation to achieve the desired density.

Generally, farmers are vague about the purpose of trees and cite edible seeds and pods, midfield shade, and pure custom (*"aquí, es la costumbre"*) to account for their presence. Indeed, the trees do have multiple uses. Seeds and pods are convenient nibbles for working farmers and are in demand in local markets. In times past bark and gum were sources of tannin and medicines, and the hard wood is still valued for construction and fence posts. But these economic products do not seem to justify intercropping trees with field crops. The small harvests of edible seeds and pods can hardly compensate for shade-lowered crop yields and obstructed plowing. The same trees often grow wild on nearby hills, offering adequate supplies of bark and gum. Instead, I suspect that over the years farmers observed that maize fields with a few trees of certain types produced good harvests. Association became custom and resulted in the present mixed farms of crops and native leguminous trees.[31]

Summary

The addition of use of organic material to soil illustrates several features of small-scale, low-energy farming. Identifying local resources and understanding their characteristics are classic examples of traditional wisdom derived empirically within particular environments. Relating these characteristics to particular soil and crop conditions is one of the highest achievements of traditional technology.

Although spatial, energy, and informational constraints typify small-scale farming, traditional farmers are uniquely able to turn these constraints to their benefit. Such practices as mulching and multiple cropping are best done on a small scale with little loss of efficiency.[32] Only simple, general-purpose farming tools are needed. In fact, some techniques such as intercropping and using midfield leguminous trees are practical only on hand- or team-worked plots. Although labor requirements are not negligible, neither are they out of range for a farmer or farm family, especially if resources are near at hand and if some of the work can be done during off-seasons, when the amount of labor demanded by regular farming operations is low.

[31]Agricultural terraces, another common traditional management technique, also help maintain field nitrogen levels. In temperate regions with about 750 mm of annual precipitation, 6 to 7 kg/ha of ammoniacal and nitrate nitrogen fall each year in rain and snow. Additions are even greater in the tropics. Terracing is one of the most effective means of reducing runoff and conserving this valuable source of nitrogen. Brady, *Nature and Properties of Soils,* p. 441; Gerald E. Schuman and Robert E. Burwell, untitled notice.

[32]National Academy of Sciences, *Making Aquatic Weeds Useful,* p. 84.

5

Soil Management Using Inorganic Amendments

If anything, silts, mucks, and sands pose greater handling problems for traditional farmers than do the bulky organic amendments discussed in Chapter 4. The weights of such materials in the quantities needed on farms are enormous, and the amount of energy required for acquisition, transport, and application strains the limited resources of small-scale, nonindustrial farmers.[1] Little wonder that traditional farmers use the power of gravity, usually in the form of running water, to move these granular field dressings whenever possible. Furthermore, unlike plant remains, silts and mucks do not lose significant volume over time, and farmers may eventually face unwanted accumulations on field surfaces. Other than the hazardous processes of erosion, there are few ways to remove such deposits without incurring high costs.

Silts

Silting to enrich soils is one of the oldest and most widespread forms of soil management in Middle America. It is distinguished here as much

[1]The opposite seems to be true in industrialized systems, where the technology for moving heavy granular material is highly advanced but where handling bulky organic materials, such as aquatic plants, is an expensive and as yet imperfect process even with the best available machinery. National Academy of Sciences, *Making Aquatic Weeds Useful*, pp. 67ff.

by its method of application as by its composition; unlike most soil amendments, silt is transferred and distributed by moving water. Given a source of silt-laden water, the problem is to control flow speeds in order to avoid either deposition or erosion in the structure that delivers the water, to induce deposition but not erosion in the fields, and to discharge surplus water without erosion. Thus, silting resolves to management of slopes and field surfaces, and farmers' tasks are primarily constructing and maintaining control forms (such as dams, canals, and terraces), leveling fields, and possibly removing excess silt. Although these projects are arduous, they cannot compare with the enormous expenditures of labor that would be required to transfer silt bodily from source to field.

Uncontrolled Flows

Perhaps the simplest form of silting is that associated with floodwater farming, where channeled or overland flows of water are slowed naturally, as on alluvial fans or aggrading river bottoms. Because flood waters also contain silts, fields are simultaneously irrigated and rejuvenated. But, except for occasionally erecting a line of rocks or brush to partially slow or spread the waters, farmers do little to modify gradients or flows. Thus, the practice is more a matter of exploiting naturally advantageous environments than of resource management.[2]

Silting fields behind silt-trap terraces was common in ancient Middle America and is still widely practiced.[3] Simple terraces of rock, earth, or plants are constructed across slopes and parallel to contours. Although their main purposes are usually to control erosion and conserve water, they also promote transfers of debris downhill. Silt accumulates where terrace ridges lower the speed of overland runoff.

Check dams *(atajadizos, presas, trincheras)* in intermittent streamways differ from silt-trap terraces in that they are built specifically to induce silting and to create level fields. Dams may be of rock, earth, or vegetable matter in any combination. By reducing the speed of channeled flows, they concentrate debris from entire catchment areas. Check dams may be built singly or in series; in some cases they form a continuous set of nearly flat fields stepped down a streamway.

[2]Julian H. Steward, "Irrigation Without Agriculture"; Kirk Bryan, "Flood-Water Farming."

[3]R. A. Donkin, *Agricultural Terracing in the Aboriginal New World;* Gene C. Wilken, "Food-Producing Systems Available to the Ancient Maya"; A.C.S. Wright, "Some Terrace Systems of the Western Hemisphere and the Pacific Islands."

Figure 5-1. Silt-trap terraces, Zinacatepec, Mexico. Curved arrow indicates flow of silt-laden water from the canal onto the *pantle* (field) through a break made in the border above a stake-and-brush dam. Seven *pantle bordos* between dam in foreground and maize field in background are indicated with small arrows.

Controlled Flows

In more advanced farming systems farmers partially control flows as well as deposition. Although they cannot control total flows of water or quantities or composition of material suspended in the water, they can decide whether and how much to silt. Thus, on alluvial fans, low diversion dams sometimes are built to spread water from relatively easily managed streams. Short ditches may direct silt-laden water to fields, which also may be checked or bordered. Although diversion dams are hardly more complex than flood-water farming or silt-trap terraces, farmers can build or break the dams and thus determine whether to flood and silt.

The ultimate in silting occurs when farmers construct dams to divert perennial or intermittent streams into canals that lead to flat or terraced fields. Thus, for centuries farmers in Mexico have enriched their fields by irrigating with silt-laden waters taken from rivers that drain eroded upstream highlands. Flooding and silting *(enlamar la tierra)* are usually done in late summer or fall, after the maize harvest. Gates are closed on permanent rock, *mampostería* (cemented rubble), or concrete dams, or temporary *estaca* (stake-and-brush) dams are built to divert river waters into canals and onto nearby fields. The problem is to design control structures that will divert and conduct silt-laden waters without slowing them enough to induce excessive deposition of debris behind the dams and in the canals.

The practice reaches even higher levels of management skill in the Tehuacán Valley of Mexico, where whole areas are organized specifically to take advantage of silt-laden waters. Around San Sebastian Zinacatepec, profitable green-maize *(elote)* farming is supported by periodic silting of fields. Small canals carrying silt-laden runoff from adjacent hills are led down along the sides of terraced fields. The low terraces are 75 to 100 m across the slope and 10 to 20 m wide, with low retaining banks *(bordos)* on the lateral and downhill sides. Brush-and-stake dams packed with rock and dirt 70 to 80 cm high and 1.5 to 2 m wide are erected 1 to 4 m below each terrace step to raise flood waters to field levels (Figure 5-1). Each flood deposits 1 or 2 cm of silt. Although expectedly low in organic matter, silts are high in plant nutrients (Table 5-1) and help maintain soil fertility under the demanding schedule of three crops per year.

The remainder of this section on controlled flows describes the most advanced silting system in Middle America, which is also found in the Tehuacán Valley, near the adjacent towns of San Gabriel Chilac and San José Miahuatlán. A major *barranca* (intermittent streamway) that descends from heavily eroded hills and valleys to the west and south fills with turbulent *broncas* (silt-laden waters) following heavy summer thunderstorms. Farmers estimate that ten to twelve storms during the May-to-September rainy season are powerful enough to fill Barranca Zapotitlán (Map 5-1).

Flood waters are captured and diverted by a series of low concrete or cemented-rubble *(cal y canto, mampostería)* dams, 1 to 1.5 m wide, 40 to 100 m long, or enough to reach across the *barranca,* and no more than 1.5 to 2 m above the stream bed. These check dams *(presas)* divert flood waters through gates *(puertos)* 5 to 8 m wide opened in the *barranca bordos.* The banks themselves are maintained to prevent breaks and flooding and are sometimes reinforced with concrete or rubble retaining walls, especially on bends and near *puertos* where erosional forces are greatest (Figure 5-2). Gates are blocked with stake, brush, and earth plugs until farmers are ready to silt. Then they are opened and flood waters flow through main channels up to 4 or 5 km long, which in turn ramify into secondary and tertiary *zanjas* that lead to individual fields.

Presa/puerto/zanja complexes are controlled by cooperative *sociedades de zanjas,* each of which has from forty or fifty to several hundred members *(socios).* Each *socio* tends his own fields and *zanjas*

Table 5-1. Composition of Silts

Sample	pH	Cond.	O.M. (%)	NO_3-N (ppm)	P (ppm)	K (ppm)	Zn (ppm)	Fe (ppm)	Lime (%)	Texture
From Zinacatepec										
D-111	8.1	5.5	2.0	13	13	473	0.3	+40.0	High	Clay
From Zapotitlán										
D-52	7.8	9.9	1.9	39	11	425	1.8	29.9	High	Silty clay
D-56	7.9	1.9	1.9	2	4	205	2.0	+40.0	High	Clay
D-112	7.9	1.1	1.9	1	8	281	1.4	+40.0	High	Silty clay

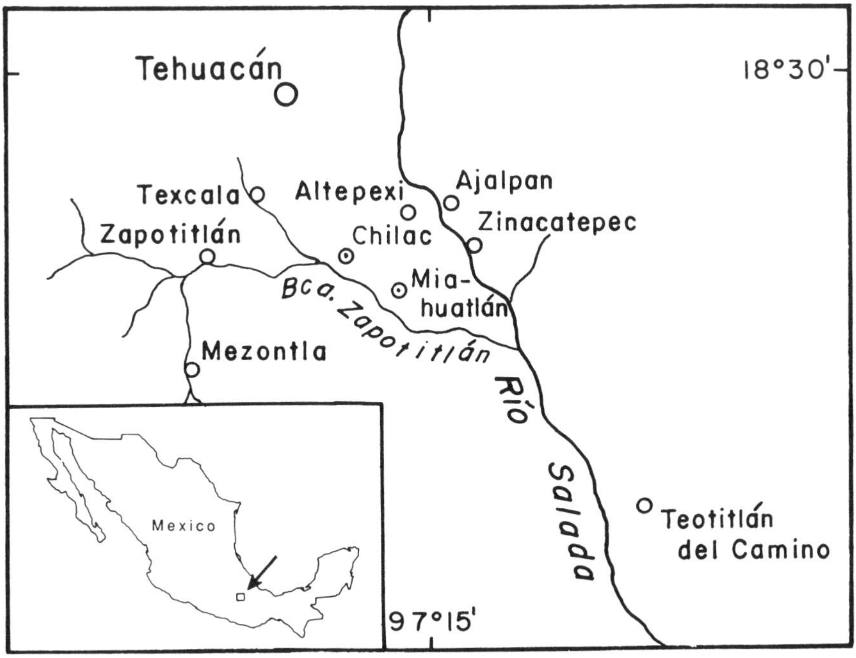

Map 5-1. Silting towns in the Tehuacán Valley and sources of silt.

and also helps maintain main channels, dams, and gates. During floods, main gates are opened, and farmers receive silt-laden water in turn: those upstream getting first rights and those downstream getting water and silt only if quantities are adequate. Usually there is enough for all who wish to flood and silt.

Fields

Individual fields *(pantles)* are essentially high-walled terraces that act as settling basins. They are variable in shape, conforming to slopes and ownership divisions. Most are quadrangular, however, and range in size from tiny plots of less than 1,000 m^2 to fields of nearly a hectare. A typical *pantle* might be 30 × 75 m, or slightly less than 0.25 ha, with more than 200 m of enclosing borders (Figure 5-3).

Pantle bordos vary in size with slope and silt accumulations. They enclose all sides of each *pantle* and are extraordinarily thick because they must control swirling flood waters. Typical *bordos* might be 3.5 to 4.5 m wide at the base, from 0.75 to 1.5 m high, and contain somewhere between 500 and 600 m^3 of material. Occasionally, huge terrace walls up to 4 m high are constructed to maintain level *pantle* beds where breaks

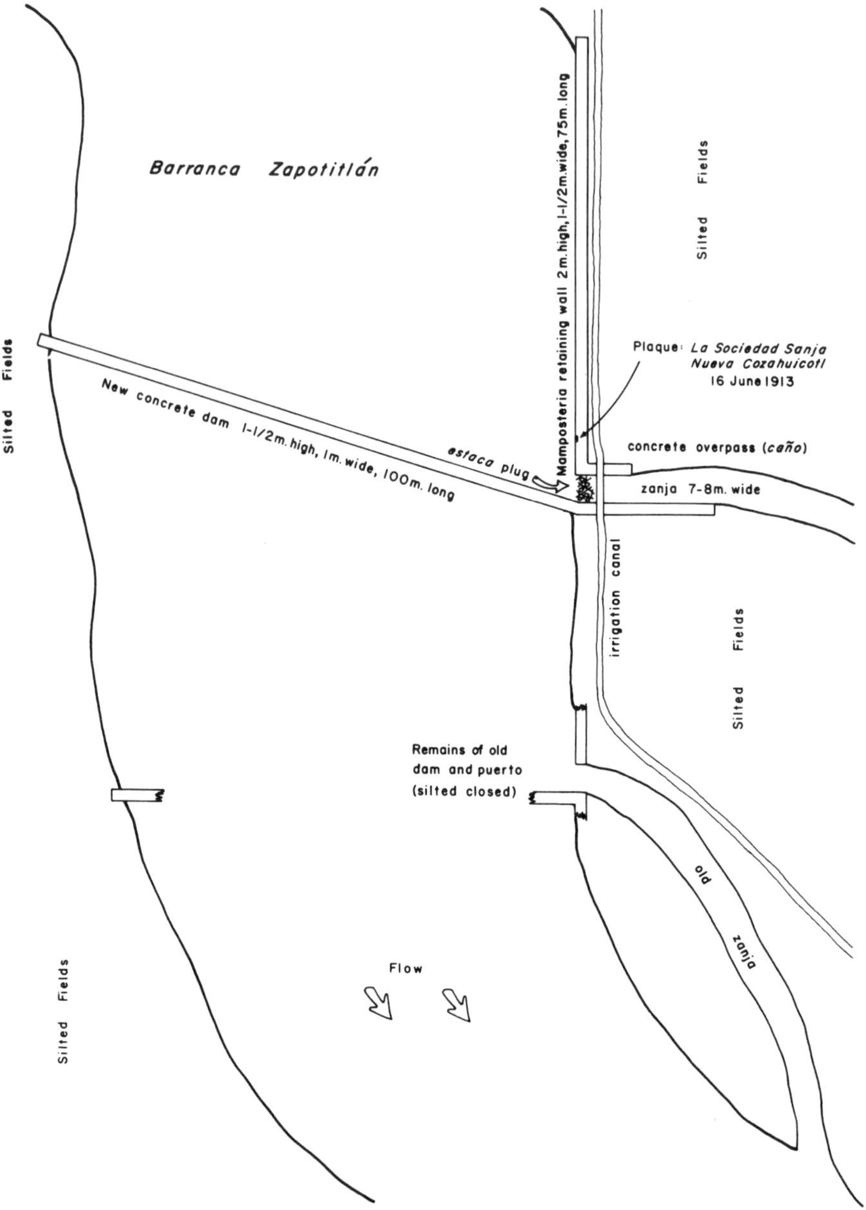

Figure 5-2. Diversion dam, gates, and *zanjas* on Barranca Zapotitlán.

Figure 5-3. Silting fields (*pantles*), San Gabriel Chilac, Tehuacán Valley, Mexico.

in slopes occur. Although of massive proportions, *pantle* walls are not protected or reinforced, and erosion is a serious problem. Flood waters quickly invade weak spots or tunnels of burrowing animals, causing bank failure and plot-surface erosion (Figure 5-4).

Silt-laden waters enter enclosed *pantles* through small gates (also *puertos*) that penetrate terrace walls at field level. The waters then discharge through outlets over sills raised perhaps 30 cm above field level. Partially ponded waters deposit a heavy layer of silt on unplanted fields (Figure 5-5). If a *pantle* has already been planted, entrance gates are opened for only a short time to allow flood irrigation and light silting.

Because flood waters do not distribute silts evenly, *pantle* surfaces must be leveled frequently. In addition, silt deposits must not build up to block irrigation water that is delivered by separate sets of channels. Periodically silts are loosened with plows, then leveled or removed with ox-drawn scrapers or in some cases bulldozers, and added to *pantle* borders (Figure 5-6).

Composition and Amount

Because summer thunderstorms are highly localized, erosion and runoff can occur within a single valley of the drainage basin of Barranca

Figure 5-4. Break in *pantle bordo.*

Zapotitlán (Map 5-1). Thus, silt may be from a single source area. For example, storms and erosion around San Antonio Texcala produce black silts of little value as soil dressing. The red clays of Los Reyes Mezontla are of pottery quality but are not particularly desirable agriculturally. Silty clays *(barros)* from Zapotitlán de Salinas are preferred and are the mainstay of farming in San Gabriel Chilac and San José Miahuatlán. Although from a single source, these silts vary in color and chemical composition (Table 5-1). Generally, however, they are richer and better textured than other *barranca* detritus.

Cooperative and individual controls along *zanja* networks allow farmers to exercise a series of decisions with respect to silting. If not enough preferred silts are present in a particular flood, main *puertos* simply are not opened, and the waters pass on downstream. Even in favorable floods the first waters tend to be debris filled and salty and are allowed to pass before gates are opened.[4] On individual fields, farmers may elect to silt or not; they can adjust quantities of water and silt simply by opening and closing *pantle* gates.

[4]Sudden cloudbursts *(aquaceros)* are not infrequent in this region. Flash floods can come quickly and with great force, trapping the unwary that have strayed into dry *barrancas* during the night before a storm. Farmers describe with some awe seeing bodies of cows, sheep, and occasionally even humans in the turbulent waters of Barranca Zapotitlán.

Figure 5-5. Silt deposition in Tehuacán *pantle*. Cracks are 7 to 15 cm deep.

Figure 5-6. Leveling silt in a *pantle*.

Opinion varies widely as to desirable amounts of silts. *Pantle* soils are dressed only lightly or not at all with animal manures or chemical fertilizers. Silt is thus not simply an amendment to existing soils; the entire growing medium is developed from fresh silts. Some farmers feel that three or four floodings (30 to 40 cm) of silt each year are necessary to maintain a strong *(fuerte)* soil and to sustain the nearly continuous alternation of maize and garlic. Others feel that 5 to 10 cm of silt are adequate. Because there is usually no shortage of silt, farmers are free to act according to their beliefs with little effort other than the management of *pantle* flooding and leveling.

Traditional Aspects

This technique of silting fits traditional patterns of minimizing costs for individual farmers. Whatever the level of silting, the quantities of material are substantial. To apply even 10 cm of silt, for example, to the typical 30-×-75-m (2,250-m^2) *pantle* involves 225 m^3 of material. Yet this material is provided at a cost of a few hours of water management and a day or two of field leveling, using only the human and animal power available to the individual farmer. The labor invested in constructing and maintaining dam and *zanja* systems cannot be ignored. But here also the contributions of each *socio* are modest, made in small increments over the years, usually during slack periods of farm work. Thus by organizing labor on a scale and at a pace congruent with the energy and scheduling constraints of traditional agriculture, the industrious farmers of Chilac and Miahuatlán have created a body of capital equipment (dams, *zanjas, pantles*) that allows them to exploit a unique resource combination of energy (gravity, flowing water) and soil amendment (silt)—a combination that is the basis of their productive and profitable farming system.

The traditional aspects of the Chilac and Miahuatlán silting technique are noteworthy in other respects. Elaborate systems of dams and gates provide decision-making opportunities at every step of the process. Within natural limits set by the timing of storms, farmers can decide when, where, and how much to silt. In so doing, they use traditional science and technology in recognizing silt types, evaluating crop and soil requirements, utilizing gravity (flowing water) for mass transfers, and devising *pantle*-management procedures.

In addition, the silt in this process is unusual as an agricultural resource because it provides the basis for social organizations that administer the resource and maintain the system. Although the organizing influence of water in traditional societies is well known, other agricultural resources generally lack the mobility of water and

PACKING SLIP

1 4 3 2 3 4 9

Via

Ship To

Alibris APEX DC 76906274-56
APEX
800 AVONDALE AVE
GRANDVIEW HEIGHTS, OH 43212-3473, US

Market ID: 76906274
Order ID: 1432349
Order Date: 09/11/2026
Shipping Method:

SKU	QTY	TITLE	TOTAL
BSM.1ATST	1	Hopeful Heart, Peaceful Mind: Managing Infertility	28.19
BSM.16GHS	1	Physics for Scientists and Engineers, Chapters 1-39	52.10
BSM.19NAM	1	Good Farmers: Traditional Agricultural Resource Management in Mexico and Central America	46.14

Order Total: $126.43

Your order has been packaged and created with great care. Enjoy! Bay State Books

#1432349

Figure 5-7. Wasted landscape near Zapotitlán, a source area for the silts of Chilac and Miahuatlán, Mexico.

thus the proclivity for accumulation, storage, distribution, and allocation that lies at the root of most water-management systems. For the Chilac and Miahuatlán silting *(zanja)* societies, however, although the link is water, the water is used less for its own sake and more as a carrier and distributor of the soil amendments that sustain the intensive, market-oriented agriculture of the region.

Source

But perhaps the most curious aspect of this silting procedure is the continued exploitation of drainage areas upstream. In poetic recognition of its value, farmers call the Zapotitlán clayey silts *"el oro de Chilac"* ("the gold of Chilac"). But what of the source region that so generously and gratuitously exports this treasure? Zapotitlán de Salinas lies in a long, narrow valley surrounded by mountains that rise abruptly from the flat to gently rolling valley floor. The entire area is dishearteningly eroded; thin sheet-washed hill soils support only sparse stands of organpipe and *nopal* (prickly pear) cactus and yucca *(palma)*. *Barrancas* that descend from the mountains continue across valley farm lands as deep gullies (Figure 5-7). Onyx-mine excavations and spoil heaps, road-material quarries, and poorly engineered drains along roadsides add to the disaster. The town, valley floor, and surrounding hills have a scabious, worn-out appearance in sharp contrast to bustling Chilac, nestled among its deep-green, carefully managed fields.

Dates on *barranca* diversion structures and local accounts suggest that the relationship between the two areas is long standing.[5] But when excessive erosion and silting began is not known. Perhaps depopulation in Zapotitlán reduced the labor available for soil and water conservation below some critical level, or emphasis on onyx mining and salt extraction reduced concern for conservation.[6] Whatever the reason, the towns and lands themselves, sleepy and wasted on one hand, busy and productive on the other, strikingly illustrate the negative and positive potentials of erosion and deposition.

Mucks

Although by composition mucks belong among the organic amendments,[7] by weight and volume they are more akin to silts. But silts are eroded and carried by streams, whereas mucks are products of deposition and require much effort to excavate and transfer from source to fields. To justify this expenditure of energy, special conditions are necessary. Mucks sources must be close to recipient fields to minimize transfer costs. Costs of muck extraction must be offset by other benefits. And usually mucks are applied only to high-value, commercial crops. Each of the cases described here meets one or more of these conditions.

Drained Fields

Southwestern Tlaxcala, Mexico, provides examples of mucking that range from rather casual practices to careful management techniques

[5]Irrigation around Chilac was well established long before the Spanish arrived. Gorgonio Gil Huerta, "History of the Foundation of the Town of San Gabriel Chilacatla." The origins of the silting process, however, are not recorded.

[6]Some springs in the area produce water with a high sodium chloride content. This water is led to evaporation terraces that form a distinctive element in the landscape and give rise to the modifying Spanish phrase *de Salinas* (salt works), which follows the town's Nahuatl name, Zapotitlán (place of the *zapotes* or *sapodillas*).

[7]Distinctions among various types of peats, mucks, and similar materials are made by Nyle C. Brady, *The Nature and Properties of Soils,* pp. 353ff., and Albert E. Wilkinson, *Muck Crops,* pp. 1ff. Rather than limiting muck to organic materials found in swampy or peaty areas, I adopt the more general definition of "partially decomposed organic matter, plant remains not discernible, accumulated in a wet place and mixed with some mineral matter." G. V. Jacks, R. Travernier, and D. H. Boalch, eds., *Multilingual Vocabulary of Soil Science,* p. 102.

that form an essential part of the farming systems. Basin-floor lands in the region suffer from high water tables and poor drainage. The situation is countered with a dense network of drainage canals *(zanjas)* that become choked with erosional debris and aquatic plants and must be cleaned periodically to restore their effectiveness.[8]

Zanja deposits are generally of two types: a sticky muck derived from *tepetate*/sand/clay soils (composition: N, 0.18 percent; P, 0.35 percent; K, 0.67 percent) and a sandy material eroded from the sands and sandy loams that develop on volcanic ash. Both types are mixed with other erosional debris and decomposed aquatic plants, and either may come directly in runoff from adjacent hills or from other fields. Farmers say that excavated muck is "rested," recognizing that the material they spread over their fields may well have been topsoil on another's plot a short time before.

Zanjas are cleaned with varying frequency: as often as two or three times a year by careful farmers; annually or even every two or three years where deposits are less plentiful or farmers less industrious. The muck is excavated by hand with shovels or spades, then either is applied immediately to field surfaces or is piled up along banks for as long as a year before spreading (Figure 5-8). Both methods have advantages. Debris stored on banks drains and settles, becoming easier to move than freshly dredged, semiliquid muck, which is usually spread over plot surfaces immediately adjacent to *zanjas*. But dried muck is difficult to spread and may form heavy clods that require pulverizing before tilling.

Sands, clays, and *tepetates* eroded from surrounding hills and mountains are distributed irregularly over the deep lacustrine deposits of the basin floor, creating a mosaic of soil types. But farmers do not have the option of matching mucks and soils because the heavy material cannot be transported over even short distances with the equipment and power available to them. Instead, they must use the mucks available in their own plot-bordering *zanjas*.

Even the decision to muck is limited. Canals must be cleaned periodically or fields will water log. Although excavated muck can be stored in piles along *zanja* banks, valuable field space is lost, and eventually successive excavations require disposal of the materials. Because of energy limitations, the only practical solution is to spread the muck over field surfaces.

Opinions vary on the value of muck *(lama)*. Some farmers feel it is of little value; they clean their drains primarily to keep them open and spread the muck over their fields as much to get rid of it as to enrich the soil. Others are convinced that *lama* is a high-quality fertilizer well

[8]Hydrologic conditions and farming methods in this region are discussed in Gene C. Wilken, "Drained-Field Agriculture."

Figure 5-8. Muck piled along sides of freshly cleaned *zanja*, Tlaxcala, Mexico.

worth the effort and apply it carefully to specialty market crops. History is on the side of this opinion. The seemingly perpetual fertility of southwestern Tlaxcala, home of dense agricultural populations since pre-Spanish times, is due in large part to periodic rejuvenation of fields by fresh or "rested" material from the drainage ditches.

The situation in Tlaxcala is a classic example of the potentials and frustrations of traditional farming. Mucking is at the heart of the system that has maintained farming in this region for centuries. But the weight and volume of material involved makes muck an extraordinarily difficult material for traditional farmers to handle. Given the opportunity, even industrious farmers turn to less-arduous practices.

At the present time, the practice of mucking hangs in the balance. New, deep drains cut by government draglines have lowered water tables and reduced the threat of field water logging. Chemical fertilizers are growing in popularity not only because more farmers can afford them but also because they are much easier to apply than traditional soil amendments. But what will the effects be on the delicate environmental balance within which the traditional system evolved? It remains to be seen whether these new solutions will have the ecological compatibility and enduring power of the old methods, which have maintained productivity in this region for centuries.

Chinampa Canals

A similar mucking technique is practiced assiduously in the adjacent Basin of Mexico, in the famous *chinampa* district, which runs from Xochimilco to Mizquic. *Chinampas* (from the Nahuatl *chinamitl*) are rectangular plots surrounded on three and commonly four sides by canals. They are constructed by excavating material from swamp and shallow lake bottoms and piling it onto plots, or by driving canals back into low-lying shore areas.[9] Plots may also be augmented with layers of aquatic vegetation and fill from shore-side locations.[10]

Like the *zanjas* of Tlaxcala, *chinampa* canals acquire deposits of eroded soils, mostly from plot surfaces, decomposed aquatic plants, and waste from villages and farmhouses. Runoff also includes residual animal manures and chemical fertilizers, with which the plots are heavily dressed. Farmers rate muck deposits on color, possibly on odor, and to a large extent on texture. The silky-fine *bien molido* (well-ground) muck is neither lumpy nor grainy, and is avidly sought for seed beds. Because canals are common property, *chinampa* farmers *(chinamperos)* can seek out deposits anywhere within their local canal network. But demand is heavy, and often they must settle for less than the best.

Analyses of samples taken from canals near the village of San Gregorio Atlapulco, in the middle of the *chinampa* district, are presented in Table 5-2. Although all samples were taken directly from canal bottoms, they may contain traces of chemical fertilizers, as noted previously. The analysis of the muck rated best *(lo mejor)* suggests the richness of this amendment.[11]

Unlike Tlaxcalan *zanja* deposits, which can be excavated directly by shovel, *chinampa* canal muck must be dredged up from deep canal bottoms with a long-handled (up to 4-m) scoop *(zoquimaitl)* and transferred to plots in canoes. As the muck is dumped into the canoes, it is carefully strained with hands and fingers to remove clay lumps and stringy plant remains. In some cases chemical fertilizers are added. The end product is a black, odorous, semiliquid soil dressing of extraordinary richness.[12]

[9]Angel Palerm distinguishes between *chinampas laguna adentro* (within a lake or swamp) and *chinampas tierra adentro* (on land). *Obras Hidráulicas Prehispánicas en el Sistema Lacustre del Valle de México*, pp. 238 and elsewhere.

[10]Pedro Armillas, "Gardens on Swamps"; Michael D. Coe, "The Chinampas of Mexico"; Elisabeth Schilling, *Die "Schwimmenden Gärten" von Xochimilco;* Robert C. West and Pedro Armillas, "Las Chinampas de México."

[11]A study of the region conducted by the Mexican government contains analyses of soils and water but not of *chinampa* mucks. C. Felipe García Castañeda, *Estudio Agrológico en la Zona Xochimilco-Chalco de la Cuenca del Valle de México.*

[12]"There is nothing better than muck, except better muck." T. B. Franklin, *A History of Agriculture.*

Table 5-2. Composition of *Chinampa* Mucks

Sample	N (%)	P (%)	K (%)	Texture	Farmers' Appraisals
D-94	0.71	0.16	0.07	Organic	Poor
D-95	0.70	0.16	0.10	Organic	Poor
D-96	0.39	0.35	2.53	Organic	Acceptable

Seed-bed quality Sample	pH	Cond.	O.M. (%)	NO_3-N (ppm)	P (ppm)	K (ppm)	Zn (ppm)	Fe (ppm)	Lime (%)	Texture	Farmers' Appraisals
D-97	6.7	3.9	13.4	3	67	+500	+10.0	+40.0	Low	Organic	Best (*lo mejor*)

Figure 5-9. Fresh muck *chinampa* seed bed, San Gregorio Atlapulco, Mexico.

Many writers suggest that muck is spread over entire *chinampa* surfaces. But dressing even a modest-size *chinampa* of, for example, 3 × 100 m with as little as one centimeter of muck would require excavating 3 m^3 with the awkward *zoquimaitl,* loading it into and out of canoes, and pouring it uniformly over 300 m^2 of plot surface. Not only is this task formidable, but it also seems unlikely that canal deposits accumulate fast enough to supply the *chinampa* district with this quantity of material. Rather, *chinamperos* dredge only enough muck to form seed beds *(almácigas).* Often these beds are laid out across the width of *chinampa* plots and are 2 to 3 m wide, or perhaps 20 to 30 m^2 in area. Low earthen borders contain the semiliquid muck, which is poured in to a depth of 3 to 5 cm for most vegetables, 8 to 10 cm for maize,[13] and even deeper for some flower stocks such as rose bushes (Figure 5-9).

The muck settles and solidifies within a few hours, after which it is cut with knives into squares with sides about equal to muck depth (Figure 5-10). Holes poked in the center of each cube with fingers or sticks receive seeds that are then covered with a light cap of pulverized

[13]The once-common practice of transplanting maize from seed beds is now rare. The expensive, multiple-cropped *chinampas* are devoted to high-value, fast-growing vegetables and flowers for markets in Mexico City.

Figure 5-10. Root cubes (*chapines*) cut from muck seed beds.

manure and soil and are carefully watered with sprinkler cans. When transplanted, the entire root cube *(chapín)* is transferred intact to the *chinampa* plot. This procedure not only transplants seedlings but also distributes the rich muck over the entire *chinampa* surface.

Whatever the particular process, it is apparent that mucking makes a vital contribution to *chinampa* fertility. Soil samples taken from *chinampa* surfaces reveal a planting medium strikingly high in plant nutrients (Table 5-3). As would be expected, the effects of surface treatments diminish with depth. But the strong impression that emerges from an inspection of *chinampa* samples is of soils whose most distinctive characteristic is management by people.

Sands

In traditional farming, sand is a popular material for modifying soil texture and adjusting field levels.[14] Sands are somewhat easier to handle

[14]A typical sample of sand used as a soil amendment in Tepetitla, Mexico, is 24.2 percent very coarse, 47.4 percent coarse, 19.2 percent medium, 8.8 percent fine, and 0.3 percent very fine.

Table 5-3. Composition of *Chinampa* Soils

Sample	pH	Cond.	O.M. (%)	NO_3-N (ppm)	P (ppm)	K (ppm)	Zn (ppm)	Fe (ppm)	Lime (%)	Sand (%)	Silt (%)	Clay (%)	Texture	meg/ 100 g CEC
D-43 (surface)	6.6	14.0	+7.0	32	+80	+500	+10.0	+40.0	Low	28	38	34	Organic (clay loam)	74.0
D-92 (surface)	8.1	15.6	+7.0	16	+80	+500	+10.0	+40.0	Low	29	36	35	Organic (clay loam)	88.0
D-93a (surface)	7.6	12.8	+7.0	31	+80	338	9.3	+40.0	Low	29	37	34	Organic (clay loam)	70.0
D-93b (depth of 0–30 cm)	8.2	13.0	+7.0	73	+80	195	4.5	+40.0	Low	29	37	34	Organic (clay loam)	84.0
D-93c (depth of 30–60 cm)	7.6	14.8	+7.0	27	37	147	1.7	+40.0	Low	27	39	34	Organic (clay loam)	74.0
D-93d (depth of 60–90 cm)	7.0	15.6	+7.0	4	38	225	1.8	+40.0	Low	30	36	34	Organic (clay loam)	74.0

Figure 5-11. Pile of truck-delivered sand ready for spreading on fields, Tepetitla, Tlaxcala, Mexico.

than other heavy amendments because they lose moisture and thus weight quickly, and, when dry, they flow in a way that aids spreading. Still, like other heavy soil amendments, sand must be available in quantity close to cultivated fields. In Middle America ephemeral stream beds or volcanic deposits, which supply sand, are often nearby. Even so, transferring sand with human- or animal-drawn containers or conveyances is so laborious and time consuming that whenever possible farmers augment their own limited labor and power sources with hired workers and trucks (Figure 5-11).

One of the most common uses of sand is to modify soil texture. Some clay soils (*fuerte,* sticky, strong) stick to plowshares when wet, form massive clods when dry, and generally inhibit field operations. Adding sand partially corrects these undesirable characteristics. Thus, in the eastern Mesa Central of Mexico, farmers apply dressings of sand 10 to 15 cm deep to loosen (*enflojar,* literally, to weaken) *fuerte* soils (Table 5-4). This amount appears to be the minimum that will noticeably affect soil structure. Yet even on the small, 1,000-m^2 plots that are common in this region, adding this much sand involves bringing in and spreading 100 to 150 m^3, or 150 to 225 m tons, a formidable task without machine assistance.

Table 5-4. Effects of Sand Treatment on *Fuerte* Soils, Tepeaca, Puebla

Sample	Sand (%)	Silt (%)	Clay (%)	Texture	meg/100 g CEC	Moisture Tension 0.1	0.3	1	15
D-33 (untreated)	51	29	20	Loam	23.6	—	25.6	19.0	14.8
D-35 (after dressings of sand)	77	14	9	Sandy loam	10.6	16.4	—	7.6	5.9

Sand is not added on any regular schedule. A difficult soil may be heavily sanded for several consecutive years, then only occasionally as needed to maintain workability. Clays from lower levels incorporated into cultivated soils by plowing and surface amendments such as muck may make a soil less workable and necessitate additional sand dressings.

As noted, without equipment and supplementary labor, the task would be more than a traditional farm family could undertake. In each of the sand-adding operations I have seen, the farmer used a small, 3- to 4-m^3 dump truck or flat-bed truck with sides to transfer the sand from source to field, and four to six hired workers *(peones)* to apply the sand. Limited data from field observations and the literature suggest that one man can shovel-load loose sand at the rate of 1 to 1.5 m^3/hr. Spreading goes much faster, perhaps 3 to 4 m^3/hr, if trucks deposit frequently in small piles. Usually only one truck is employed so that the same workers load and spread. Thus, loading and spreading are discontinuous, and allowance for travel time must be made. An overall procurement, transfer, and application rate of 0.5 m^3/hr seems reasonable and generally in the range of other estimates.[15] Thus, to dress a 1,000 m^2 plot with 10 cm of sand would require 200 man-hours plus costs of a truck and driver.

Adding sand to soil to lower salt levels costs even more than adding sand to improve texture. In most sands downward percolation is rapid, whereas upward capillary movement is limited.[16] This factor makes sand a desirable surface material in areas with high water tables, where salts tend to accumulate on the surface.[17] If salt has already formed, farmers may cover it with as much as 30 cm of sand. It is usually not necessary to dress an entire field because salt formation is not uniform; it tends to be heaviest in low spots, which is fortunate. To apply a 30-cm layer of sand to a 1,000-m^2 plot would require 600 man-hours of labor and associated truck costs—a heavy burden for most traditional farmers even if field reclamation and increased crop yields warranted the expense.[18]

[15]J. Müller, "Labour-Intensive Methods in Low-Cost Road Construction"; *World Bank Study of the Substitution of Labor and Equipment in Civil Construction,* Tech. Memo. 18, "Spreading Activities in Civil Construction."

[16]Brady, *Nature and Properties of Soils,* pp. 184ff.; Food and Agriculture Organization, *Sandy Soils,* especially papers by J. P. Baudelaire, "Irrigation of Sandy Soils," and P. J. Dielman, "Drainage of Sandy Soils."

[17]Food and Agriculture Organization and United Nations Educational, Scientific, and Cultural Organization, *International Source-Book on Irrigation and Drainage of Arid Lands in Relation to Salinity and Alkalinity,* especially chap. 13, "Reclamation of Saline and Alkali Soils."

[18]I have found no Middle American parallel to the ancient European practice of applying salt to fields as a selective herbicide. Allan E. Smith and D. M. Secoy, "Salt as a Pesticide, Manure, and Seed Steep."

Figure 5-12. Surface salt accumulations and reduced grain crop.

Because covering salty surfaces with sand is feasible only for small areas, widespread salt formation (*salitre, tequesquite,* from the Nahuatl *tequixquitl*) in fields is one of the most intractable problems faced by traditional farmers.[19] Leaching is possible where adequate surface water is available and drainage systems are in place. But if these conditions were met, salt formation probably would not occur in the first place. If soils are only slightly salty, salt-tolerant crops such as alfalfa, beets, carrots, and garlic may thrive. But, generally, eliminating severe or widespread salt formation exceeds the technical and financial resources of traditional farmers, and such conditions generally result in lowered yields and abandoned fields (Figure 5-12).

Chemical Fertilizers

The convenience and potency of chemical fertilizers appeal strongly to traditional farmers. Adequate quantities are easily transported from market to village and from homestead to field. Application can be precise both as to quantity and location. Traditional farmers custom-

[19]The problem is not new in Middle America. Bernardino de Sahagún mentions several saline soils recognized in Aztec Mexico. *General History of the Things of New Spain,* pt. 13, pp. 254–55.

arily measure out exact amounts of fertilizer in small containers or handfuls and carefully apply it directly around plants or to seed beds or mix it with other soil conditioners.

Several deterrents limit use. The main one is cost, especially in subsistence farming, where cost/benefit relationships are not clear. Supply can also be a problem because local markets may not be able to guarantee delivery at critical times, usually at planting or early emergence. Furthermore, unlike information about locally developed amendments, information on chemical fertilizers is exogenous. Data from manufacturers or extension agents are often incomplete, especially with respect to local soils, climates, and crops. Results have been predictable: many losses and even near disasters, and slow acceptance as farmers gradually elaborate information from external sources with local experience.

But, despite obstacles, chemical fertilizers are steadily securing a permanent place in traditional soil management, especially in commercial operations. Many farmers by now have had considerable experience with the various nitrogen, phosphorus, and potassium combinations and with secondary and micronutrient responses, and have developed their own guidelines for application. Often these practices differ substantially from manufacturers' recommendations. Although perhaps inadequate for optimal responses, the quantities applied are a reasonable compromise between economic constraints and anticipated results.

In some cases traditional attitudes help avoid certain negative aspects of chemical fertilizers. For example, assigning the characteristics of "hot" and "cold" is widespread in Middle America, most commonly to bodily condition and foods,[20] but also to other phenomena, including soils. For many farmers chemical fertilizers are "cold," unlike "hot" animal or green manures. They believe that to achieve balance the soil should be fed organic materials to offset the "cold" chemical fertilizers. This belief encourages continuance of mulching and manuring, and thus maintenance of soil body and tilth.

Summary

Soil science is often treated as a purely technical topic, amenable to examination without reference to cultural contexts. This approach may

[20]The "hot/cold" belief is a modified form of Greek humoral pathology transmitted through the Arab world and Spain to the New World. Richard L. Currier, "The Hot-Cold Syndrome and Symbolic Balance in Mexican and Spanish-American Folk Medicine"; George M. Foster, "Relationships Between Spanish and Spanish-American Folk Medicine"; John M. Ingham, "On Mexican Folk Medicine."

be valid as long as the discussion is limited to the physical and biological aspects of soil/plant relationships. But because soil classification and management are bodies of knowledge and technology associated not only with biophysical farming environments but also with the cultures that utilize them, they escape technical confines.

Thus, despite the technical information provided, this chapter and the previous two essentially deal with a cultural question: given the universal biophysical qualities of soils and plants, what can farmers do if they must operate within the traditional constraints of local knowledge, materials from the immediate environment, and power derived from other than fossil fuel? The answer is, "quite a bit." Traditional understanding of soil qualities, especially for agricultural uses, is considerable. Both in classification systems and management methods, traditional approaches reveal an effective grasp of soil characteristics and potentials. Equally important, through long and intimate association with particular environments farmers have learned to recognize a variety of local materials that enhance the physical and chemical properties of soils. And, finally, they have developed low-cost methods for transferring bulky or heavy amendments from sources to fields.

Many specific examples of traditional soil management are impressive. Who would not admire the elaborateness of Almolongan soil classification? The effectiveness of leaf mold and canal muck as soil conditioners? The complexity of Tehuacán silting networks? But we must go beyond admiration for specific practices and view the whole as a functioning system that is based on detailed understanding of soil characteristics and plant requirements and that relies almost entirely on local resources.

6

Slope Management

The mountainous regions of Middle America are products of primordial forces. Still-active volcanos and frequent earthquakes have created a jumble of abrupt mountains, folded hills, and steep escarpments cut by deep canyons and narrow valleys. This precipitous natural landscape contrasts sharply with the cultural landscape. Over the years traditional farmers, hardly less a primordial force, have modified thousands of hectares to permit effective use of the land.

This discussion of surface geometry is divided into two parts: slope management (covered in this chapter) and field-surface management (see Chapter 7). The distinctions are of duration, scale, and purpose. Under slope management I group relatively permanent changes made to whole fields or even entire hillsides primarily to control runoff or irrigation water. Included in such efforts are terracing and leveling. In contrast, field-surface adjustments are usually temporary, lasting no more than a season or year, and are designed to improve biological, climatic, or edaphic conditions for crop plants. They take such forms as ridges, pits, raised beds, or even individual plant mounds. The distinctions are not sharp. Field-surface features may cover a field or hillside and affect as much of the surface as permanent slope systems. Conversely, many slope-control measures have profound effects on soil nutrients and moisture levels. But a rough grouping into categories is useful, especially when reviewing labor requirements for construction and maintenance.

More than one writer has commented on the curious lack of scientific attention to agricultural landforms.[1] The few studies on the Western Hemisphere deal mostly with the pre-European period; surveys of contemporary slope or field forms are rare, and rarer still is information on the costs and benefits of restructuring the natural landscape.[2] This review does not correct these deficiencies but instead focuses on a few common features found in Middle America for which some data are available.

Slopes represent a dynamic process. On natural slopes the forces that move materials downhill are balanced by factors that resist downward movement. The impact of gravity, the fundamental degradational (erosive) force, can be seen directly in land movements such as in soil creep or slides, and indirectly in the flow of water. In the case of surface runoff, effects increase with distance. Not only does water accelerate as it travels downhill, but, in addition, sheet wash is quickly concentrated into small channels. These degradational forces are resisted by such factors as soil shear and erodibility, and vegetation.[3]

Most slope modification is aimed at controlling surface flows, either runoff or irrigation water.[4] Factors that influence surface flows include rainfall amounts and intensities, soil structure and texture, vegetation, and large and small surface features. Other than protect the surface against splash erosion, farmers can do little about rainfall. The physical characteristics of soils can be altered to a certain extent by adding organic materials or sand to improve infiltration rates or water-holding capacities. Although a complete, permanent vegetative cover is effective for controlling erosion, it is inimicable to the needs of most farming. Most important is slope management, by which the gradients over which water develops erosive power are modified.

[1]Berl Golomb and Herbert M. Eder, "Landforms Made by Man."

[2]Although focused on the pre-Spanish period, R. A. Donkin's *Agricultural Terracing in the Aboriginal New World* contains considerable information on contemporary terracing. See also J. E. Spencer and G. A. Hale, "The Origin, Nature, and Distribution of Agricultural Terracing"; A.C.S. Wright, "Some Terrace Systems of the Western Hemisphere and Pacific Islands." For a recent survey and analysis, especially of the socioeconomic aspects of hill farming, see Andrés R. Novoa B. and Joshua L. Posner, eds., *Agricultura de Ladera en América Tropical.*

[3]This concept is developed by M. A. Carson and M. J. Kirkby, *Hillslope Form and Process,* chaps. 3 and 4. The term *erodibility* includes several factors, such as particle size and permeability, that affect soil response to degradational forces. W. H. Wischmeier and J. V. Mannering, "Relation of Soil Properties to Its Erodibility."

[4]Some writers argue that moderate erosion is beneficial to crop production because it exposes fresh mineral soils. For example, Wright ("Some Terrace Systems") maintains that stabilizing slopes with terraces in the humid tropics would nullify natural soil-rejuvenation processes.

Although slope management has the control of water as its primary purpose, other benefits result from it. For example, several slope-control structures act as settling basins: check dams create level fields in streamways; surfaces of the Tehuacán *pantles* (fields) are periodically dressed with silt. Thus, by reducing gradients (slope management) and water flows, farmers increase percolation or allow irrigation (water management) and reduce erosion or induce deposition (soil management). In addition, improved fields permit other practices (mulching, manuring, fertilizing) and tolerate intercropping or multiple cropping (space management). In the final analysis, it is difficult, and perhaps pointless, to decide which purpose is paramount.

Individual decisions to modify surface geometry rest ultimately on farmers' perceptions of feasibility and on anticipated returns in relation to costs. In some cases favorable benefit/cost ratios are apparent. For example, the substantial yields from irrigated fields usually justify the costs of leveling. Less apparent are the returns from erosion control. It is testimony to the perspicacity of traditional farmers that so many are willing to devote so much effort to controlling degradational forces.

Irrespective of scale, most slope- or topographic-control measures can be reduced to the several phases of materials handling: excavation or acquisition, transfer, and deposition. But each phase requires substantial work, which in traditional farming systems must come from humans or animals. The two basic solutions to the problem are to combine sources of power or to reduce scale. For rapid, large-scale slope modification, massing human and animal power may be attractive, especially where labor is abundant. But slope control lacks the organizing tendencies of water management, and institutional requirements limit this solution to those societies that can control labor.[5] Reducing scale is the more common solution. Most traditional slope projects can be constructed and operated by a family or small group. Even so, over long periods of time the cumulative results of small-scale slope-control projects can be impressive.[6]

[5]Organization, supervision, and incentives are critical elements in large-scale labor-intensive projects. Basil Coukis, *Labor-Based Construction Programs;* World Bank, *World Bank Study of the Substitution of Labor and Equipment in Civil Construction.* Contemporary land-leveling and terracing projects in the People's Republic of China offer spectacular examples of what can be done with massed labor in a highly organized society.

[6]Many Old World terrace systems are products of centuries, even millennia, of effort. For example, the famous terraces of northern Luzon were most likely built over a period of perhaps 2,000 years. H. O. Beyer, "The Origin and History of the Philippine Rice Terraces." See also, Harold C. Conklin, *Ethnographic Atlas of Ifuqao.* Paul Wheatley ("Agricultural Terracing") argues that, despite a long period of development, some terrace systems in Vietnam are the result of preconceived and carefully executed plans.

Unable to concentrate power from either mechanical equipment or massed labor, traditional farmers accept that reshaping the agricultural landscape requires patience. Carefully scheduled stages and predetermined completion dates characterize projects in industrialized systems. In traditional slope management extensive terracing or erosion control is undertaken with little thought as to the pace of work or when it will be finished.[7]

In addition, slope modification is never permanent. Powerful degradational forces are inescapable elements in hillside environments. Once natural slope stability is disrupted, farmers must continuously manage hillside ecosystems to ensure that exposed soils and control structures remain intact. Slopes are destabilized not just by natural forces but also by most farming practices. For example, original clearing and subsequent cultivation reduce the impeding effects of vegetation on surface flows and soil erosion. Tillage loosens soils and weakens shear resistance. Successful hill farming therefore requires compensatory practices that decrease or interrupt overland flows or reduce gradients. But entire hills cannot be leveled, and if gradients are reduced on some parts, inevitably they are steepened on others. Thus, slope modification itself creates unstable conditions.

With these brief introductory remarks, we turn to a discussion of specific slope treatments. The list is not comprehensive, and it is not arranged chronologically. Although simpler forms precede more complex structures in the discussion, it cannot be assumed that they did so in history. Simple methods of slope control are still widely practiced in Middle America and serve admirably where they are appropriate to environmental and social conditions.

Check Dams

Check dams provide an excellent illustration of the interplay between natural forces and human management. Check dams *(atajadizos, bordos, presas, teceras, terrazas, trincheras)* are built in intermittent streamways *(arroyos)* in arid and semiarid regions where sudden storms and sparse vegetation combine to produce heavy erosion. The dams are constructed across smaller streams where the force of water rushing down after a storm will not destroy the rock, earth, and branch structures. Check dams reduce the speed of the flowing waters and thus their ability to carry eroded materials. As the water slows, suspended debris is deposited and forms flat, flood-irrigated and subirrigated fields behind each dam.

[7]Gene C. Wilken, "Traditional Slope Management."

Figure 6-1. Check dams (arrows) and depositional fields, southern Tlaxcala, Mexico.

Since the small dams can slow but not completely contain heavier stream flows, they must provide for overflow. This also opens possibilities for a sequence of dams, each checking the flow and capturing some of the suspended debris. In some *arroyos* check dams form continuous steps with the top of each downstream dam level with the base of the one above it (Figure 6-1).

Check dams are often built in stages. In this respect they display the process-rather-than-project approach of so many traditional management activities. A narrative account from an Otomí informant in Hidalgo, Mexico, offers insights into the process:[8]

> *Atajadizo* is the name of fields made in gullies. It is also the name used for the actual wall that is constructed to retain the water and the soil. . . . *Atajadizos* need to be strong in order to withstand the force of the water when it rushes down the gully. An *atajadizo* that is well built has a double wall: the outside wall is pitched in order to hold back the water; the inside wall holds the soil that accumulates. It also has a floodgate *(compuerta)* so that the extra water can spill over. Furthermore, the wall itself is curved in order to deflect the force of the water.

[8]Kirsten Johnson Haring, *Preliminary Report on Field Work Carried Out in Hidalgo, Mexico Between October 1973 and April 1974,* pp. 19–22.

It may be necessary to use mortar and it is best to use the largest solid rocks possible.

An *atajadizo* isn't built all at once. Usually a farmer starts with a low wall across the path of an *arroyo*. It takes a few years until the water has brought down enough debris *(basura)* and soil to level with the top of the wall. Then, the farmer will build up the wall a bit more, and so on, little by little until s(he) has built up a tall strong wall and a large level field. A well-made *atajadizo* is level so that the trapped water will cover all parts of the field evenly. It may be necessary to level the field by hand and, also, to tear down parts of the gully in order to enlarge the field. But this is only the case when the gully is very narrow and uneven. In most cases the flow of water takes care of enlarging and leveling out a field.

A well-made *atajadizo* always has a wall that is higher than the field behind it. This is necessary because water must be trapped so that it can soak into the field. But if the field is at the same level as the wall, the water will just flow over it and waste.

There is no need to fertilize an *atajadizo* because every rainy season the water brings down new debris and soil.

A number of variables, such as gully configuration, force of water, materials availability, and type of construction, affect the amount of labor required to build a check dam. Erasmus[9] found that during a 5-hr workday[10] one man had no trouble excavating 7 m^3 of earth with a shovel or 3.5 m^3 of rock with a steel crowbar.[11] Comparable information on the time required to place these materials in a check dam or similar structure is not available. To build single- or double-fitted stone walls of the types described here probably would take no more than twice the time spent excavating. Thus, as a first approximation, a total time (excavating plus placing) of 1 to 2 m^3 per working day seems reasonable.

But suitable materials are not always near at hand. Erasmus establishes that during a working day a man can travel a total of perhaps 10 to 12 km carrying 20 to 25 kg of earth or rock (and return with only the container) for a daily transfer capacity of 200 to 300 kg/km (Table 6-1). The volume of materials delivered to a construction site declines sharply with distance, as the table shows. Transferring bulk materials horizontally or vertically is one of the most difficult tasks faced by low-energy societies. The data here, for example, indicate that rock transfers consume as many man-hours as the excavation itself when transfer distances exceed 100 m.[12]

[9]Charles J. Erasmus, "Monument Building."

[10]A 5-hr or, more commonly, 6-hr day is reasonable for sustained, hard work. It is also a common workday in rural Middle America.

[11]A figure of 2.7 m^3 of rock per day for manual excavation from a quarry in Africa includes loading into trucks. J. Müller, "Labour-Intensive Methods in Low-Cost Road Construction."

[12]A World Bank study recommends a 30-m limit haul distance to loads carried by humans. Coukis, *Labor-Based Construction Programs,* pp. 190–92.

Table 6-1. Materials-Transfer Rates by Human Carriers

Trip Distance, One Way (m)	Material	Total Trips, 6-hr Day	Total Distance, One Way (km)	Total Weight Carried (kg)	Average Weight Carried per Trip (kg)	Total Volume Carried[a] (m^3)	Performance Index, kg/km/Day, One Way
50	Earth	234	11.7	4,716	20.2	3.63	236
100	Earth	134	13.4	2,672	19.9	2.06	267
250	Rock	39	9.8	1,088	27.9	0.76	272
500	Rock	23	11.5	575	25.0	0.40	288
750	Rock	17	12.8	586	34.5	0.41	440
1,000	Rock	13	13.0	295	22.7	0.20	295

SOURCE: After Erasmus, "Monument Building," pp. 284–87. Rock data adjusted from 5- to 6-hr day.

[a]Volumes calculated: 1 m^3 = 1,300 kg of earth or 1,440 kg of rock.

Check-dam dimensions vary enormously. A good set of measurements comes from Chihuahua, Mexico, where prehistoric *trincheras* ranged from a few centimeters to 3.5 m high, 1 to 165 m long, and 15 cm to 3 m thick.[13] Representative *trincheras* 0.5 to 1 m high, 6 to 9 m long, and 0.33 to 0.5 m thick would contain from 1 to 7 m^3 of material. The ancient builders probably did not spend much time transferring rocks because there appears to have been abundant suitable material in this volcanic region, especially in the eroded gullies where the check dams were built. I estimate that a 3- to 4-m^3 structure could be erected in perhaps 2 to 3 man-days.

Mortared check dams are rare, and I have no data on their labor costs. A few figures from other structures, including stone-faced road drains and *mampostería* (cemented-rubble) foundations and walls suggest that mortaring a fitted rock wall can be done in the general range of 0.25 to 1 m^3/man-day, depending on availability and suitability of materials and quality of work.

The calculations of labor invested in check-dam construction are based on limited data and should be considered only rough approximations. Nevertheless, it appears that even the elaborate structures are relatively inexpensive to build. Once the dams are in place, field development proceeds with little additional effort; silt deposition is induced by reducing stream speed, and the fertile fields essentially create themselves.[14] The same reduction of stream force helps control gully erosion. Because the periodically silted and wetted fields are highly productive, I suspect that check-dam farmers enjoy a high rate of return on their labor investment. Certainly on the semiarid Mesa Central of Mexico benefits from check dams must more than justify costs. In the arid north they are one of the few means by which farming can be supported.

Check dams do not directly affect general slope characteristics. But they and their associated alluvial fields are in precarious balance with stream forces. Water topping a dam or pouring through a leak can cause rapid, drastic erosion. Although some of the rock *trincheras* in Chihuahua are still intact after more than 500 years, most are susceptible to failure, especially during sudden, heavy stream flows. Frequent maintenance is necessary to prevent breaking and undercutting, and subsequent rapid erosion of deposited soils. In this regard check dams share a characteristic of all terraces. Although gradients, and thus

[13]Laurance C. Herold, *Trincheras and Physical Environment Along the Río Galiván, Chihuahua, Mexico,* pp. 90ff; William A. Howard and Thomas M. Griffiths, *Trinchera Distribution in the Sierra Madre Occidental, Mexico.*

[14]Ernst Griffin and Howard W. Dennis, "A Mexican Corporate Campaign in Conservation"; Comisión de La Malinche (Mexico), *Realizaciones en la Montaña de La Malinche.*

Figure 6-2. Debris in shifting-cultivation plot retards runoff and erosion.

debris-transport forces, are reduced on their beds, gradients on terrace faces are increased, making them highly vulnerable to erosion.

I have seen or read reports of check dams in the Mexican states of Chiapas, Chihuahua, Hidalgo, Mexico, Oaxaca, and Tlaxcala, and I suspect that they are actually found in most states and countries in Middle America. They are not simply artifacts of primitive slope management. Modern government erosion-control projects in Chihuahua and Tlaxcala include rock check dams identical to those constructed for centuries by traditional farmers.[15]

Sloping Terraces

The most extensive slope-management systems in Middle America consist of sloping terraces, on which original gradients are changed

[15]A somewhat analogous procedure is followed by Zuñi farmers, who set out rows of brush to induce deposition of wind-blown sand and dust. Frank Hamilton Cushing, *Zuñi Breadstuff*, pp. 165–66. Some central Asian people similarly create level areas of wind-deposited materials that are then flooded. Food and Agriculture Organization and United Nations Educational, Scientific, and Cultural Organization, *International Source-Book on Irrigation and Drainage of Arid Lands in Relation to Salinity and Alkalinity*, p. 510.

only slightly or not at all. In their simplest form, sloping terraces consist of nothing more than rows of logs or rocks laid perpendicular to slope gradients to partially check down-slope surface wash (Figure 6-2). In their more elaborate forms, earth or rock embankments follow hill contours. At their fullest development, sloping terraces have carefully shaped earth-and-rock or rock embankments *(bordos)* reinforced with economic plants and parallel drainage ditches *(zanjas).*

The main functions of sloping terraces are to control erosion and accumulate moisture. Although slope angles are not deliberately changed, most sloping terraces develop modified gradients through natural processes (over time, deposition builds up behind embankments) or human intervention (in team-worked fields consistently turning furrows downhill transfers soil in that direction and reduces gradients).

There are many variations of sloping terraces in Middle America. West describes the distinctly central Mexican *metepantle,* in which low embankments are paralleled immediately down slope by drains, forming a *zanja/bordo* combination that is probably the most popular slope-control technique on the Mesa Central.[16] Characteristically grain is planted in the fields and *maguey* (*Agave* spp.) on the *bordos* to stabilize them and also to produce flower-stalk sap (*aguamiel,* from which *pulque* is fermented) and many other products.[17] Other economic plants commonly found on *metepantle* embankments are *mesquite* (*Prosopis* spp.) and the native fruit trees *tejocote (Crataegus mexicana)* and *capulín (Prunus capuli).* West identifies *metepantles* as a form of semiterracing or incipient terracing. But their long history and extensive modern use suggest that they are not a rudimentary stage of terracing but rather a well-developed technique uniquely adapted to the physical and agricultural conditions of central Mexico.

Despite their antiquity and extensive modern use, the mechanics of *zanja/bordo* systems and their relationships to physical and biological conditions are not well understood. We do know that *bordos* act as hillside check dams that prevent surface wash from accelerating and channeling. Although some deposition occurs behind the embankments, their purpose seems more to reduce erosion and absorb moisture than to trap silt. The reinforcing *maguey* is used because, although it suffers in saturated soils, it fares well on the higher *bordos. Zanjas* act as small field reservoirs to catch seepage and any ponded runoff that tops the *bordos.* Although some *zanjas* are excavated up slope from the

[16]Robert C. West, "Population Densities and Agricultural Practices in Pre-Columbian Mexico, with Emphasis on Semi-Terracing."

[17]In addition to *pulque,* the plant provides fiber *(ixtle),* building materials, roof thatching, fuel, and foods (including young leaves and edible larvae). Gene C. Wilken, "The Ecology of Gathering in a Mexican Farming Region."

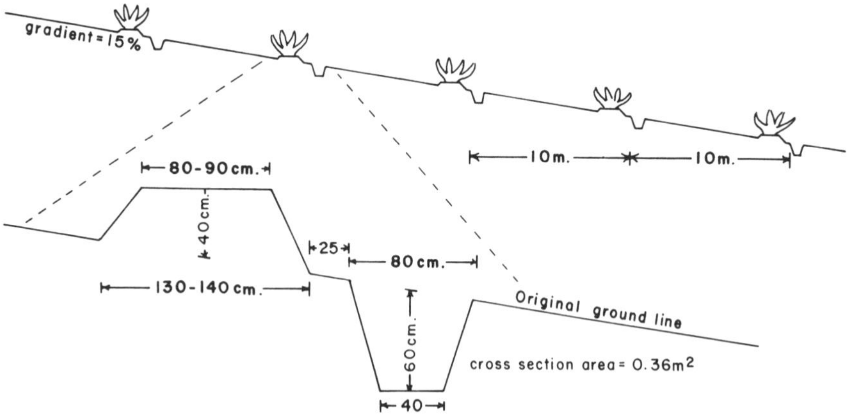

Figure 6-3. Sloping (*zanja/bordo*) terrace.

embankments, most are below, where they are less likely to fill with sediments. In porous soils water stored in *zanjas* gradually percolates into field substrates.

Sloping terraces are a method of controlling slopes at low cost, a feature appreciated by traditional farmers and modern planners alike. In 1972 an agency of the Mexican government stated that "the salvation of rainfall *[temporal]* agriculture consists in the modification of traditional systems of cultivation" and enthusiastically undertook a series of conservation projects involving *trincheras,* stone-faced terraces, and *metepantles.*[18] One such operation in western Tlaxcala provides an opportunity to corroborate field observations with government data on labor inputs in *zanja/bordo* projects.

Farming is marginal in this region. Average rainfall of 600 to 700 mm is barely adequate for grain crops, and below-normal years are hazardous. Rains come mostly in the period from May to September, often in intense, convective showers. Much of the region is gently rolling to hilly. The thin, sandy soils have little water-holding capacity and are easily eroded. These problems are common in central highland Mexico, as are sloping terraces as a solution.

With the help of government engineers, farmers and *ejiditarios*[19] laid out lines of drains along contours. Trapazoidal *zanjas,* 80 cm wide at the top sloping to 40 cm at the bottom, were cut 60 cm deep, for a

[18]Dirección General de Conservación del Suelo y Agua (Mexico), *Resumen Gráfico de Tabajos de Control de la Erosión de los Suelos y de Conservación del Agua de las Lluvias,* p. 14. See also Secretaría de Obras Públicos (Mexico), "Obras a Mano."

[19]*Ejiditarios* are those who received expropriated lands *(ejidos)* under the Mexican land-reform program.

Figure 6-4. Sloping terraces with *maguey* in *zanja/bordo* system.

cross-section of 0.36 m^2. Dividing strips 50 to 60 cm wide but only 30 to 40 cm high were left every few meters to stop movement of water in the drains. Excavated material was piled immediately up slope from the drains in geometrical *bordos* 40 cm high and 80 to 90 cm wide at the top sloping to 130 to 140 cm at the bottom (Figure 6-3). The *bordos* were not compacted except by incidental foot traffic, which is why they have a larger volume than the *zanjas* from which they came. After *zanja/bordo* construction was complete, young *maguey* was planted along embankments at 3-m intervals (Figures 6-4 and 6-5).[20] Except for the engineers' transits, the only tools used were shovels and spades, and picks when consolidated *tepetate* layers were encountered.

In addition to providing engineering assistance, the government paid workers on the basis of *tareas* (or *journales*), the fair or normal amount of work that can be done in 6 hr.[21] *Tareas* are based on type of work, terrain, and material. Table 6-2 lists representative *tareas* for this project. Within these ranges precise amounts are determined by specific

[20]Field preparations for transplanted *maguey* are reviewed by Candido Cruz López, "Estudio Agrológico Regional del Estado de Tlaxcala," especially no. 3, pp. 123ff., "El Maguey y su Cultivo"; Miguel Macedo Enciso, *Manual del Magueyero,* pp. 7–28.

[21]Workers proceed at their own pace. Energetic individuals may complete their *tareas* in 4 or 5 hr, while slower workers might take 7 or 8 hr.

Figure 6-5. Planting young *maguey* on *bordo* above down-slope *zanja*.

Table 6-2. *Tareas* for *Zanja/Bordo* Construction

Material	*Tarea* (m, linear)	Approximate m^3 (cross-section = 0.36 m^2)
Soft soil (*blando*)	20–25	7–9
Moderately compacted (*duro*)	10–15	3.5–5.5
Tepetate	5–10	2–3.5
Rocky (*rocoso*)	1–5	0.5–2

conditions of a particular site. For example, a worker may be expected to excavate only 5 or 6 m (linear) of *zanja* in solid *tepetate* but 10 to 12 m if the *tepetate* layer accounts for only the bottom few centimeters of a trench in otherwise loamy soil. The data on *tareas* were checked against individual farmers' estimates elsewhere in the region and against quotas set for voluntary community work *(comunidad)* in villages.[22] Correspondence was excellent. Farmers working on their own

[22]Many villages in Mexico adhere to the ancient custom of *comunidad,* in which one day each week is devoted to community projects. All adult males not otherwise occupied are expected to participate.

lands *(dueños)* are more motivated than hired workers *(peones)* and will exceed the work rates given here. But for *peones* the figures appear to be good indicators of what one man can do in a day on a sustained basis.

Those unfamiliar with hand labor may be surprised at the amount of material that one man can move and thus the relatively low cost per cubic meter. Government engineers assigned to these projects were delighted with results and soon were boasting that, given favorable conditions (moderate terrain, little rocky material), they could build a variety of public works better and cheaper with hand labor than with machinery.

To estimate labor requirements, we need to know the width of *metepantles,* or distance between *bordos;* this distance depends on slope gradient and soil texture and depth. Cruz López proposes the following general guide for *maguey* plantations: on gentle slopes (1 to 5 percent), *metepantle* width is generally 5.0 to 6.5 m; on moderate slopes (5 to 10 percent), it is 4.0 to 5.0 m; and on steeper slopes (10 to 15 percent), it is 2.5 m.[23] In western Tlaxcala, where perhaps the absorptive sandy soils reduce surface runoff, intervals are somewhat greater. Width of *metepantles* on the government-assisted erosion-control project ranged from 5 to 20 m. With examples mostly from nearby Apan, Hildalgo, West suggests a range of 12 to 30 m on moderate (5 to 10 percent) slopes down to 3 m on steeper ones (up to 25 percent).[24]

Although far from complete, the data permit some preliminary analyses of labor requirements for *zanja/bordo* slope-control projects. An average terrace width of 10 m will serve for the example. If we allow an additional 10 percent for contour sinuosities, a hectare would have a total of 1,100 m (linear) of *zanjas* and *bordos,* for a total volume of ($0.36\ m^2 \times 1{,}100 =$) $396\ m^3$ of material to be excavated and formed into embankments. Total labor requirements would vary from 44 man-days in soft or sandy soils to more than 100 man-days in hard *tepetate.* Generally this type of work is not undertaken on rocky slopes or in pure *tepetate.* Therefore, most projects probably fall into a range of 60 to 70 man-days/ha.

These estimates are for direct labor only; they do not include time spent planning, administering, surveying, or supervising. On large communal or *ejido* projects, perhaps 20 percent more time is needed for these tasks. In individual small-farm operations the time required for planning, laying out, and supervising is inseparable from time spent during normal farm work and constructing terraces.

With the data presented here, it would be relatively easy to estimate labor costs for a *zanja/bordo* project in almost any type of material. But

[23]Cruz López, "Estudio Agrológico," no. 3, pp. 129–30.
[24]West, "Population Densities and Agricultural Practices."

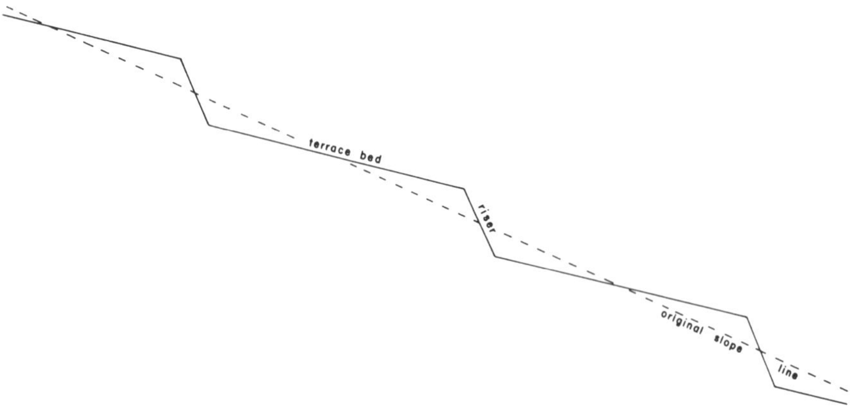

Figure 6-6. Chamula sloping terrace (not to scale).

Figure 6-7. Sod-faced Chamula terraces, Chiapas, Mexico.

on individual small farms slope control is undertaken by the farmer and his family, perhaps with a few friends on an exchange-labor basis, and an occasional hired hand. The cost is more realistically represented by the 60 or 70 days it takes to treat a hectare of hill land than by a peso amount. As one Otomí farmer put it, "*bordos* don't cost money; they cost work";[25] time and effort can be spread conveniently over indefinite periods so that *bordo* construction becomes more of an ongoing process than a project.

A different approach to sloping terraces is found on rolling limestone hills near San Cristobal de las Casas in the Chiapas highlands of southern Mexico. The Chamula terraces (after the Chamula Indians of that region) typically are 5 to 10 m wide with relatively steep risers 0.2 to 1.0 high (Figures 6-6 and 6-7). Variations include rock- or sod-faced risers, small riser-edge *bordos,* and heel drains. Modifications of slope gradients are achieved by normal crop cultivation and down-slope debris transport. Unlike the team-plowed fields of central Mexico, Chamula terraces are hand cultivated with broad hoes *(azadones).* Crops such as maize, beans, potatoes, and wheat are popular. Sheep graze on grass covers that develop during fallow periods.

Not infrequently, combined forms of sloping terraces have multiple functions. For example, sloping fields in the rugged Sierra Norte del Puebla, Mexico, are bounded by *zanjas* and meter-high rock walls (Figure 6-8). Debris transport could be controlled with much lower structures. But the walls also serve as convenient depositories for rocks collected from fields and as barriers to the often carelessly herded sheep of that region.

Although sloping terraces are not as unstable as flat (bench) terraces, they are subject to accelerated erosion if not tended. West notes that "on hillsides where *metepantles* have been abandoned and where *maguey* rows have not been maintained, severe sheetwash and gullying have destroyed most of the thin soil cover."[26] Water ponded behind an embankment will pour through a break with concentrated force causing severe local erosion (Figure 6-9). Thus, like all terrace forms, sloping terraces partially destabilize hill-slope and water relationships and create potentially hazardous conditions.

So far, the discussion has treated slope as a negative element to be reduced or controlled. But sloping fields also provide some advantages, such as easily worked soils, good drainage, and the assistance of gravity in certain field tasks. On flat lands on the floor of the Puebla-Tlaxcala basin of central Mexico, high water tables and poor internal drainage make for difficult farming conditions. In addition to cutting drains,

[25]Quoted in Haring, *Preliminary Report on Field Work,* p. 23.
[26]West, "Population Densities and Agricultural Practices," p. 367.

Figure 6-8. Sloping terraces with *zanjas* and high rock walls, Sierra Norte del Puebla, Mexico.

Figure 6-9. Break in *bordo* and gully erosion.

Figure 6-10. Curved fields, Tlaxcala, Mexico.

farmers create slight gradients by curving field surfaces convexly (Figure 6-10). Plowing creates this field camber. Except for passes to open small, central drains, furrows are turned toward field centers; plot surfaces thus are gradually worked into a curved form in which field centers are perhaps 20 cm higher than the edges. Once plots have been shaped, only routine plowing is required to maintain the curved form. The process is almost exactly opposite that which occurs on sloping terraces, where down-slope debris transport and transfers from plowing reduce gradients.

Flat Terraces

The ultimate in slope management is reached in flat, or bench, terraces, with which traditional farmers transform sloping hillsides into level fields.[27] An enormous range of types and subtypes is found in Middle America, which has had a terrace tradition for hundreds, perhaps thousands, of years. Some bench terraces are simply constructed, carved out of hillsides with broad hoes and periodically abandoned. Others are

[27]Gene C. Wilken, "Drained-Field Agriculture."

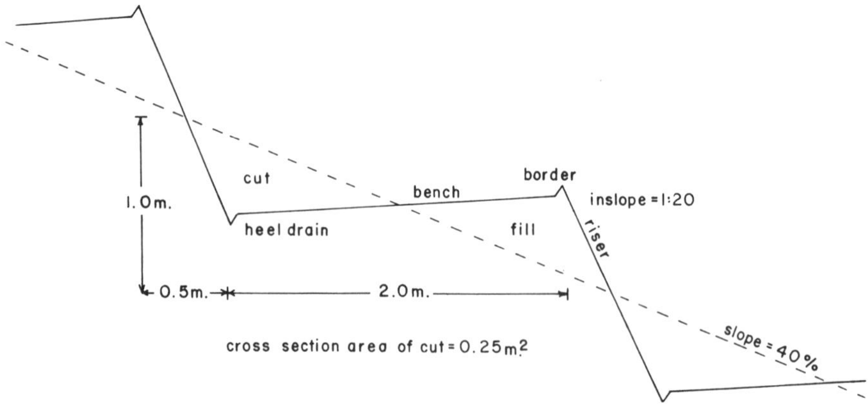

Figure 6-11. Flat, or bench, terrace.

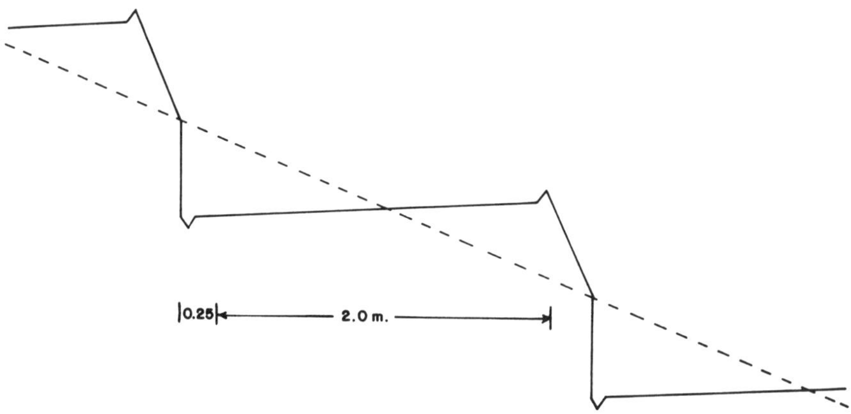

Figure 6-12. Bench terrace with vertical and sloped risers.

carefully shaped and improved over the years. Only a few details of construction and maintenance are discussed here.

Typically, bench-terrace risers, or faces, are tilted back as little as possible to maximize bed width while still maintaining some measure of stability (Figure 6-11). Benches on *temporal* terraces have slight inslopes so that surface runoff flows toward the heels, often into small drains, rather than over easily eroded risers. Irrigated versions are similar except that beds are absolutely flat. Small ridges (again *bordos*) along riser edges, rather than inslope gradients, keep water from overflowing. Ditches at the heels are larger than on *temporal* terraces to carry irrigation water.

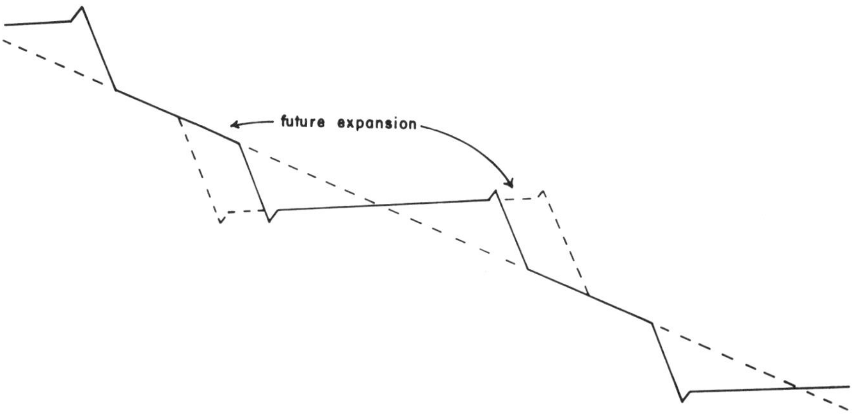

Figure 6-13. Bench terraces with unterraced space between tiers.

A variation on basic bench-terrace design suggests one way to enlarge planting surfaces without sacrificing stability. Figure 6-12 shows a terrace with the riser sloped back 0.5:1 in the fill material but with the compact, cut face left vertical. Bed width is increased by more than 10 percent with this technique.

Farmers need to take several factors into account when designing flat terraces. Slope is the primary determinant of terrace configuration. As a general rule, the steeper the slope, the narrower the bench or bed. In addition to surface geometry, however, soil types and depths and working design are important. For example, underlying bedrock limits the depth of terrace cuts. But even if soil depths are adequate, deep cuts introduce infertile subsoils onto plot surfaces. In some cases farmers leave space between terrace tiers and gradually expand them over the years so that less-desirable material is only slowly incorporated into planting beds (Figure 6-13).

In addition, farmers need to consider costs. The basic objective of bench terraces is to produce the degree of levelness needed to control surface water. Unlike check-dam fields and *metepantles,* bench terraces are not leveled to an acceptable degree by natural forces. Direct labor investments are necessary for almost all phases of construction. One way to reduce these costs is to build narrow terraces, which are less expensive than wide ones. The volume of material that must be excavated and transferred (cut and filled) to terrace a given area varies geometrically with terrace cross-section, whereas the number of terrace tiers varies arithmetically with bench width. Thus, if N terraces of bed width BW and riser height RH involve excavating and filling V m^3 of material, $2N$ terraces of dimensions $0.5BW$ and $0.5RH$ will produce the same amount of level planting beds but require only $0.5V$ m³ of cutting and filling. Little wonder that traditional farmers often opt for narrow bench terraces even where slopes and soils would permit wider fields. With no need to accommodate cumbersome powered equipment,[28] traditional farmers can keep bench terraces narrow and labor investments correspondingly small. In the final analysis traditional terrace dimensions are based on slope angle, soil type and depth, and working-space requirements.

Although labor costs can be reduced by building narrow terraces, they are still considerable because traditional bench terraces are usually handmade. Common tools include shovels, spades, broad hoes, pickaxes, and crowbars. Only a few animal-drawn implements, such as buck *(fresno)* or drag scrapers, are useful. Although some farmers have simple bubble levels, many merely estimate slopes and angles, and use small amounts of running water to establish suitable gradients for heel drains. New terraces usually are built by parties of several men (Figure 6-14).

Labor costs are subject to two sets of variables. Bench terraces are constructed in many forms and sizes, and are found on a wide range of slopes that comprise many soil types and depths. Equally important are variations in the human labor units; individual workers differ greatly in physical and emotional states and in responses to environmental conditions (temperature, humidity). These elements are not unique to terrace construction but hold true for all estimates of production rates involving humans or animals. In the example here data distilled from a number of field observations and supported by farmers' estimates and government work quotas are applied to a typical bench terrace to illustrate construction costs.

On moderate slopes of 40 percent, benches 2 m wide are not unusual, with backslopes on the order of 0.5:1 (Figure 6-12). The cross-sectional area of such a terrace is 0.25 m². With soft *(blando)* soil, an experienced workman should be able to excavate, fill, and form small drains in about 20 to 25 m (linear) (5.0 to 6.25 m³) in a 6-hr day. About thirty-seven tiers of terraces of this size are needed to cover 100 m of slope. Allowing an additional 10 percent of terrace length for contour sinuosity, each hillside hectare would require 4,070 m (linear) of terrace, or a volume of material slightly more than 1,000 m³. At the indicated rates, a three-man work party could produce a hectare of terracing in 54

[28]Although bench terraces are not unique to traditional farming, they are not generally recommended for industrialized systems. Thus, Glen O. Schwab et al. (*Elementary Soil and Water Engineering,* p. 119) say: "The original bench terraces . . . were costly to construct and were not always well adapted to modern cultivation equipment." See also Peter Crossley and John Kilgour, *Small Farm Mechanization for Developing Countries,* pp. 14–15; Soil Conservation Service, *A Manual on Conservation of Soil and Water,* pp. 96ff.

Figure 6-14. Bench-terrace construction, Quezaltenango, Guatemala.

to 68 days, for a total of 163 to 204 man-days. In the process, they would create more than 8,000 m^2 of flat fields on what had been sloping hillside. Although bench terraces 2 m wide are not unusual, narrower terraces suffice, especially for manual-cultivation systems, and they are cheaper to build, as explained previously.

In heavier soils and clays construction rates drop to 10 to 15 m (linear)/man-day, and in rocky soils or *tepetate* to 5 to 10 m. Work rates are lower than those developed for *zanja/bordo* construction (Table 6-2), because greater care is required to shape and compact terrace fills and to form drains and borders on flat terraces. Thus, total labor investments could easily run to more than a man-year per hectare.

Maintenance of flat terraces adds to these costs. If the ultimate in traditional slope management is reached with flat terraces, so too is the ultimate in slope artificiality and instability. Maximizing terrace-bed

Figure 6-15. Extensive *tablón* systems, Totonicapán, Guatemala.

Figure 6-16. Severely eroded old terrace system near Amozoc, Puebla, Mexico.

Figure 6-17. Rock-walled bench terraces, central Mexico.

space requires that risers be as steep as possible and, consequently, highly susceptible to degradation. On terraces that cover entire hillsides, forming interdependent systems (Figure 6-15), failure in one part unleashes concentrated degradational forces that can devastate whole sections down slope (Figure 6-16).

In addition, flat terrace beds themselves are only provisionally stable because down-slope movements of soil and water are controlled only as long as risers remain intact and heel drains perform satisfactorily. Constant surveillance and maintenance are necessary continuing costs of bench terraces.

Stability can be increased by adding artificial walls, but such additions increase costs as well. Where soils are compact, risers may simply be exposed subsoil, perhaps with a protective cover of vegetation to impede erosion. But in looser soils terraces may need artificial facing such as rocks or poles, in some cases secured with mortar or lashings (Figure 6-17). The ultimate in defense is a separate structure such as a cemented-stone or even concrete wall embedded in the hillside against the slope.

Terrace-wall construction is costly, especially if materials must be brought from a distance. In addition, terrace walls are not built in increments, as are other traditional slope-control structures. Most terrace walls appear to have been built all at once, perhaps because of the need to complete a certain lineal distance or to establish proper base/top/height proportions during original construction. Consequently, the traditional farmer contemplating construction of terrace walls must be in a position to commit large amounts of time and effort to the project over a relatively short period of time.[29]

There are thus two basic reasons for the labor costs incurred in the construction of flat terraces. First, most flat terraces offer few opportunities for gravity to assist with cutting, filling, or leveling, which require much labor. Second, costs are considerable because walls are often necessary. Flat terraces must offer significant benefits—either the production of high-value crops or the creation of fields in extraordinarily space-scarce circumstances—to justify such heavy investments of labor.

Unirrigated *Tablones*

In western Guatemala narrow unirrigated hoe terraces, or *tablones,* are common wherever hillsides exceed 20 or 25 percent and are found on slopes as steep as 70 percent. Basically they are bench terraces. But their unique operation and solution to the problem of field preparation merit special attention.

Tablón dimensions do not differ greatly from those of conventional bench terraces. Flat or slightly convex beds 0.5 to 1.5 m wide rise in tiers 0.2 to 0.8 m high. Slope angle partially controls terrace design as it does in basic bench structures; generally terraces on slopes steeper than 40 or 45 percent have benches less than one meter wide, whereas on more gentle slopes they range from 1.0 to 1.5 m. Riser backslopes also vary with gradient, from near vertical on gentle slopes to 0.5:1 on steep hillsides. Wheat is the most common crop grown on *tablones,* although other *temporal* crops such as maize and broad beans *(Vicia faba)* are popular too.

But terrace design is determined also by cultivation practices. The narrow *tablones* are formed and cultivated solely with *azadones,* the universal farming tool of this region. A big 29 cm (11.5 in.) wide blade fitted to a 120 cm long wooden handle is the most popular size. Although *tablones* occasionally are broken down and reworked, by and large they are permanent forms that remain through many crop cycles and even though extended fallow periods.

[29]Of the spectacular granite vine terraces of Minho Province, northern Portugal, Dan Stanislawski says: "In the case of the low [3-4-ft] walls, the owner is probably of modest circumstances and without much capital to invest. Many small farmers of the Minho build terraces little by little on 'their own time,' that is, after the necessary work of the day is done." *Landscapes of Bacchus,* p. 38.

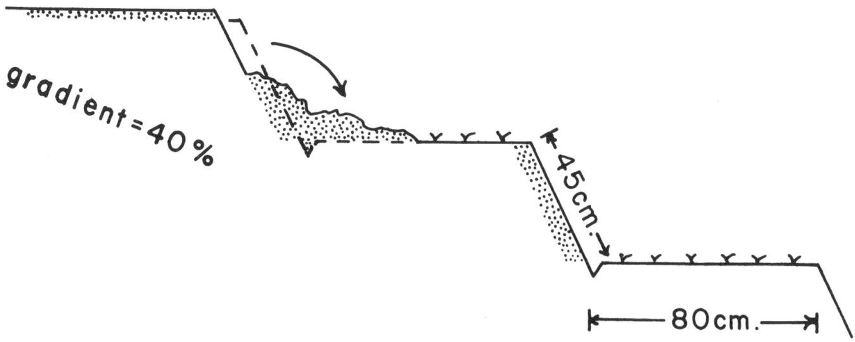

Figure 6-18. *Tablón* tillage.

A description of *tablón* cultivation will clarify the process. Prior to planting, terrace beds are not tilled; instead, risers are shaved off and the excavated material is spread over benches to establish a planting surface. Work begins at the top of the slope. Farmers, often standing on the next lower tier, swing *azadones* down to cut off an appropriate slice from the old terrace face. Swing arcs reestablish backslope angles on the newly exposed risers. Two or three swings usually are enough to cut through the risers and recut the 5 to 10 cm deep heel drains. The resulting soil is pushed and pulled across the bench with the *azadón* to bury the old surface under a layer of loosened soil (Figures 6-18 and 6-19).

For wheat, farmers strive for a layer about 5 cm deep. Thus, in Figure 6-18 about 9 cm would be cut from the riser. As soon as a few tiers have been prepared, wheat is sown and immediately covered by the same process: riser faces are again shaved and trimmed with *azadones,* and loosened soil is extracted from heels and drains to cover *(tapar)* the seed with about 2 cm of soil. Thus, a total of about 7 cm of soil is spread over the *tablón* surface. A slightly convex field surface is often developed at this stage by raking with *azadones.*

This method of plot preparation, in which old planting surfaces are buried undisturbed complete with stubble or burned residue (*basura,* literally, garbage) and root systems from previous crops, has several advantages. First, erosion potentials are greatly reduced because the basic terrace structure remains intact and only a thin layer of loose soil is exposed to splash and sheet erosion. Second, organic materials and ash are interred at root levels of the new crop. Third, layers of loosened soil may surpress weed growth. Fourth, soil from the whole riser profile is added to planting beds each year, and thus all soils are rotated on a long-term basis. For example, it will take 6 to 7 yr to work through the 0.8-m terrace illustrated in Figure 6-18 with 12.5-cm cuts each year.

Figure 6-19. Tilling *tablones.* Loose soil for planting bed secured by shaving riser with *azadón.*

Fifth, and perhaps most important, the system requires much less effort than would digging up terrace beds. With steady work, farmers expect to till, sow, and cover a *cuerda* (approximately 440 m^2) of *tablón* in a 6-hr working day, considerably more than would be possible by ordinary shovel or spade tillage.

Several features of *tablón* design make it possible to conserve labor in this way. Beds are narrow, permitting the entire operation of cutting and spreading to be done from one position. In addition, backslope angles and heel drains are reformed as part of seed preparation, not as separate operations. But most important, arduous soil lifting—an essential phase of plot preparation when using shovels or spades—is replaced by gravity-assisted downward transfers of soil, and farmers can move along rapidly without excessive physical exertion. One man-day per *cuerda* (or 23 man-days/ha) may not seem an impressive rate for field preparation. Yet in speed and efficiency, the *tablón* system must rank high compared with other methods of working steep slopes.

Irrigated *Tablones*

The irrigated *tablones* of western highland Guatemala have little in common with their dry namesakes. They are small (perhaps 10 to 30 m

Figure 6-20. High-walled, hand-irrigated *tablones*, Sololá, Guatemala.

long, 2 to 6 m wide) and rise in steps 0.5 to 2.0 m high up steep valley sides in Cakchiquel Indian areas of the department of Sololá.[30] Irrigation water led into heel canals is splashed over plot surfaces with small basins *(palanganas)*. Riser *bordos* are not needed because the plots are neither flood nor furrow irrigated. Surfaces are completely leveled, however, to prevent ponding of splash-applied or rain water (Figure 6-20). To prepare fields, riser faces are not shaved; instead, plot surfaces are tilled with *azadones* at the same time that leaf mold *(broza)*, manure, and field debris are worked into the soil. With complete control over slope, water, and soil conditions, farmers harvest three or four crops of vegetables each year, most of which are sold in local, regional, or national markets.[31]

[30]Kent Mathewson *(Irrigation Horticulture in Highland Guatemala)* presents a detailed review of irrigated *tablón* operations in the same general region. He reports (pp. 82–83) construction of one 3 × 30 m *tablón* per *tarea*, which is lower than the rates developed here for unirrigated *tablones*. However, his example includes substantial soil conditioning as well as careful plot formation in preparation for at least two years of farming with splash irrigation.

[31]T. David Johnson, "Análisis de Actividades Necesarias para la Producción de 11 Especies Horticolas (Hortalizas)"; William C. Merrill, Lehman B. Fletcher, and Michael S. Hanrahan, "Vegetable Production and Marketing in Guatemala."

Figure 6-21. Rock-faced *tablones,* Quezaltenango, Guatemala.

Equally hard-worked *tablones* cover the narrow Samala River flood plain at Zunil, Quezaltenango. Some fields occupy relatively flat land; others ascend gentle slopes as terraces. Plots are tiny, no more than 10 m wide and less than 25 m long. Like the Sololá terraces, they are hand cultivated with broad hoes and splash irrigated, although with long-handled scoops *(palas)* rather than basins. Most Zunil *tablones* are rock faced, either dry wall or mortared. Terraces are completely flat to accommodate splash irrigation and are bordered by an elaborate system of small canals so that all parts of the plots are within range of scooped irrigation water (Figure 6-21).

Discontinuous Control Structures

If crops or exposed soil are not continuous, the control structures do not need to be continuous either. In Middle America several types of discontinuous or partial terraces are used. Circular pit terraces *(cepas)* around individual trees in hillside orchards are common and illustrate the technique.

Cepas are constructed by cutting and filling just as are ordinary flat terraces (Figure 6-22). They perform all customary terrace functions

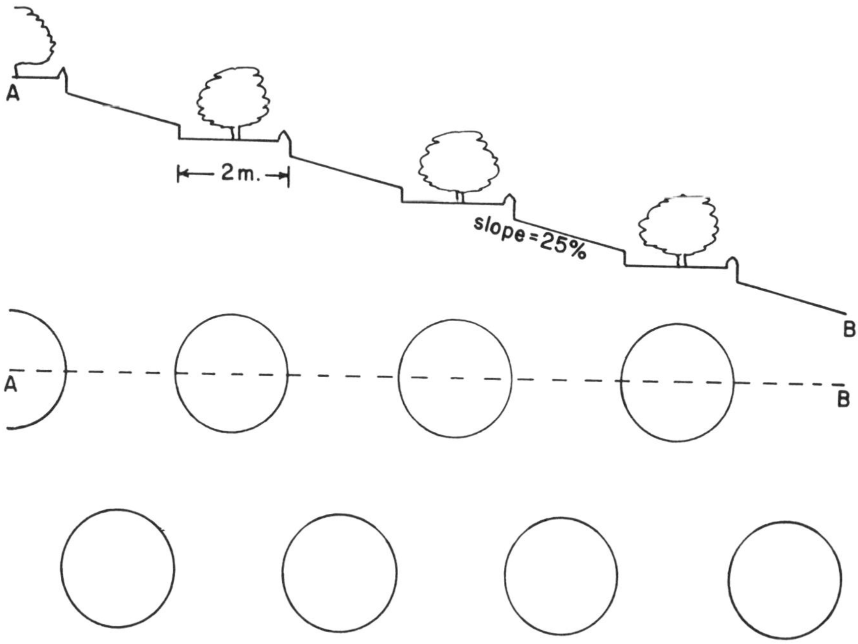

Figure 6-22. Pit terraces (*cepas*).

such as controlling erosion, conserving moisture, and facilitating cultivation. But, in addition, this form of terracing has distinct advantages. Unworked portions of slopes remain at original gradients under vegetation, thus reducing erosion potentials. And irrigation water delivered through small canals or tubes is confined to immediate tree areas, a great saving over flood irrigation (Figures 6-22 and 6-23).[32]

But most important, because reorganization of surface geometry is kept to a minimum, little labor is necessary. For example, constructing one of the *cepas* shown in Figure 6-22 would require excavating less than 0.2 m^3. At indicated spacings, each hectare would have about 922 *cepas*. Assuming construction rates similar to those for bench terraces, it would take only 23 to 58 man-days to construct pit terraces in a hectare of hillside orchard.

The process in new orchards is to level circular areas only slightly larger than seedling crowns. As trees grow, terraces are enlarged. Such phased expansion suggests the process rather than project approach.

[32]The benefits of individual tree-sized terraces in connection with irrigation by team- or tractor-drawn water tanks are reviewed by Leonardo Obregón Formoso, *Desarrollo de la Fruticultura por una Sistema Práctico y Económico de Riego en Zonas Aridas y Semi-Aridas.*

Figure 6-23. *Cepas* newly planted with avocado seedlings pock a hillside, Veracruz, Mexico.

Materials transfers from cut to fill are down slope, assisted by gravity. Once formed, *cepas* trap surface wash and debris to moisten and enrich the soil around the trees. Thus, the several principles of traditional slope control find expression in this single management form.

Summary

Reshaping natural landscapes into forms suitable for farming poses special problems for nonindustrial systems. Lacking the concentrated energy of fossil fuels, traditional farmers are limited to human and animal power for transferring volumes of material and for controlling forces that transport debris.

Animals are an important source of power, especially for horizontal transfers such as land leveling. But for many operations, such as lifting and fine excavating, animal-powered equipment is unsuitable.[33] Furthermore, to use animals efficiently may require larger plot sizes than

[33]As far as I know, there are no animal-powered soil-lifting implements that would be equivalent to back hoes or trenchers.

are feasible. For example, slope and soil conditions may not permit construction of bench terraces wide enough to accommodate animal-drawn equipment.

Massed labor offers another solution. If adequate labor supplies and appropriate sociopolitical institutions exist, most slope-control projects can be accomplished completely or substantially with human labor. But the organizations and customs necessary make this solution available to only a few contemporary societies.

Thus, power available to traditional farmers for slope management is generally limited to relatively small amounts of human labor. These constraints give rise to several distinct characteristics of traditional slope management. First, slope-control projects are generally of modest size. Although large projects may be attempted by social units such as villages or *ejidos,* most practices are scaled to families or small groups.

Second, because materials transfers pose special problems, distances are kept as short as possible. Transfers across a few meters can be accomplished without the use of containers or conveyances. But longer distances require loading and unloading, and volumes moved per unit of time decline rapidly. Traditional farmers avoid long-distance transfers or resort to powered equipment whenever possible.[34]

Third, for bench-terrace construction, beds are kept narrow to minimize the amount of material that must be excavated and transferred. Initial labor investment and working-space requirements probably are as important as slope and soil conditions in determining bench-terrace dimensions.

Fourth, gravity is used whenever possible. Check-dam fields are created almost entirely by stream-delivered and stream-deposited sediments. Gradients on sloping terraces are modified in large part by the work of surface wash. The ingenious *tablón* farmers of western Guatemala work with, not against, gravity in tilling and planting their narrow wheat terraces. Some of these management practices would be impractical if materials had to be transferred by humans or animals.

Fifth, traditional slope-management activities are processes more than projects. Because the supply of labor is severely limited, significant slope modification is accomplished in small increments. Schedules and completion dates are rare, and construction may continue for years, even generations.

[34]Prehispanic construction work in Middle America may have involved relatively long-distance transfers of earth and rock by human carriers. Angel Palerm, *Obras Hidráulicas Prehispánicas en el Sistema Lacustre del Valle de México;* Angel Palerm and Eric Wolf, *Agricultura y Civilización en Mesoamerica.* But the long lines of basket-laden workers found on some modern Chinese and Indian projects have no parallel in contemporary Middle America.

Sixth, slope management requires much labor for construction and continuing labor for maintenance. Slope control modifies original gradients and creates long-term hazardous conditions. Structural failure and environmental degradation are more likely to result from reduced rather than increased labor inputs.[35] Moreover, once modified, gradients and surface runoff patterns cannot be quickly reversed. Slope management should therefore be undertaken with extreme caution if there is a possibility that labor for maintenance may not be available in the future.

Despite these constraints and hazards, traditional farmers are capable of impressive achievements in slope management. Over the years thousands of hectares of mountainous Middle America have been molded into forms that effectively conserve moisture, control erosion, and create productive fields in areas otherwise marginal or unsuitable for farming. These achievements, as well as the occasional failures, merit our close attention.

[35]Thus, I do not agree that dense or increasing population is necessarily correlated with accelerated erosion, as is suggested, for example, by Sherburne F. Cook, *Soil Erosion and Population in Central Mexico.* More often population decline or sociopolitical disruption leads to neglect of resource management.

7

Field-Surface Management

Perhaps no other activity so characterizes traditional farming as the careful shaping of field surfaces into features that improve crop plant environments. Using plows, shovels, hoes, and sometimes bare hands, farmers laboriously mold a wide variety of mounds, ridges, pits, and beds to modify edaphic and microclimatic conditions on surfaces and in root zones. In the process they create distinctive, often striking plot surfaces that have no counterpart in the machine-worked fields of industrialized agriculture.

Farmers make microtopographic adjustments in their field surfaces in order to gain benefits they perceive for particular crops in particular environments. Local custom prescribes the specific forms that surface adjustments take; most farming regions have two or three standard forms which in turn are interpreted or modified by individual farmers. The result is a nearly infinite array of minor field forms and variations that defy comprehensive survey. This chapter describes a few common minor field forms used in Middle America, reviews their functions (especially as understood by the farmers who create them), and estimates labor costs in man-hours or man-days.

Shaping loose earth into mounds, ridges, or raised planting beds is the most common way of altering surface microtopography. Such mounding usually begins when fields are first prepared for planting and may be modified in subsequent cultivations. Rearranging planting surfaces varies from simple mounding around individual plants or plant clusters to carefully constructing elaborate planting beds.

Except for maize mounds, which will be discussed separately, raised forms have these primary purposes: to provide beds of loosened earth for easy planting and subsequent root or tuber development; to improve drainage and lift plants above flood or furrow irrigation water on flat fields, or to retain moisture on slopes; to facilitate cultivation and weed control; and to mix together residual or applied organic materials, manures, or chemical fertilizers, or all three. Modifying surface geometry produces other results, including altered field microclimates. But the reasons listed are those most often given by traditional farmers in Middle America for reshaping field surfaces.

Nonirrigated Fields

Raised Beds

Raised planting beds are found in an enormously wide range of environmental and cultural circumstances, suggesting that the benefits in relation to costs are varied and high.[1] From the many examples of large and small raised beds in Middle America, the striking *camellones*[2] of western highland Guatemala will illustrate the formation and operation of raised beds here. Their use is so widespread that the term *camellón* or *tablón* quickly identifies rounded or flat-topped raised beds almost anywhere in the region.

Soils are first loosened and cleaned of weeds with broad hoes *(azadones),* then carefully shaped into flat-topped beds about 150 cm wide and 30 cm high (Figures 7-1 and 7-2). Young farmers boast that if not hindered by matted root systems of invading grasses or weeds, one man can form a *cuerda* (about 440 m^2) of *camellones* in one day. Usually men work in groups of three or four so that *camellón* formation proceeds rapidly in the tiny fields of that region.[3] On flat fields work

[1]In addition to being used in traditional Africa, Asia, and Latin America, raised beds are also found in such contemporary systems as the French Intensive Method, which itself has roots in ancient European agriculture. John Jeavons, *How to Grow More Vegetables,* pp. 1-5.

[2]The term *camellón* is not specific to raised planting beds but is applied to many types of linear field forms and even to ridges of sand and gravel deposited along roadsides for road repair.

[3]One-third of the farms in the region are smaller than 0.7 ha. Escuela Facultativa de C. C. Económicas de Occidente y Comité Interamericano de Desarrollo Agrícola, *Tenencia de la Tierra y Desarrollo Socio-Económico del Sector Agrícola en Guatemala,* p. 175. Labor-intensive *camellón* farming is found only on smaller, hand-worked plots.

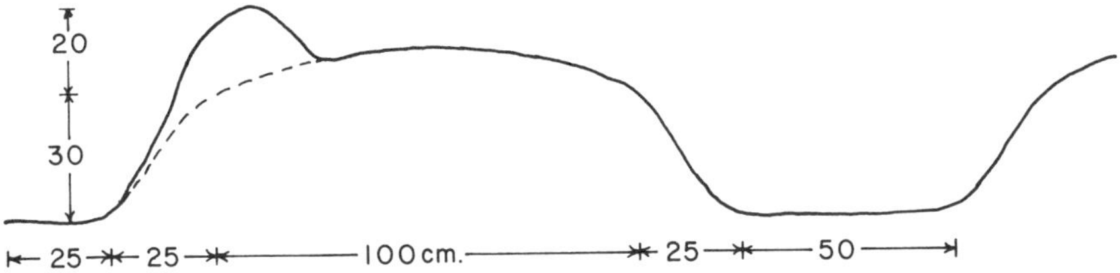

Figure 7-1. Cross-section of typical *camellón* of western Guatemala. Single maize mound (*mul*) shown to illustrate position and size.

Figure 7-2. Young maize on *camellones,* Quezaltenango, Guatemala.

begins from either side with careful attention to spacing, alignment, and general shape. On hillsides farmers first form narrow downhill ridges perpendicular to the slope. Then, after carefully checking general alignment, they fill in the up-slope portions (Figure 7-3).

Despite deep furrows that develop between the raised beds, *camellones* are not irrigated. According to the farmers who make them, their purposes are to provide a superior planting bed, loosened soil for improved root development, and a field surface that traps and makes effective use of precipitation yet maintains a well-aerated root zone.

Camellones are not formed on gradients exceeding 20 to 25 percent. As gradients steepen *camellones* may evolve into the flat hoe terraces

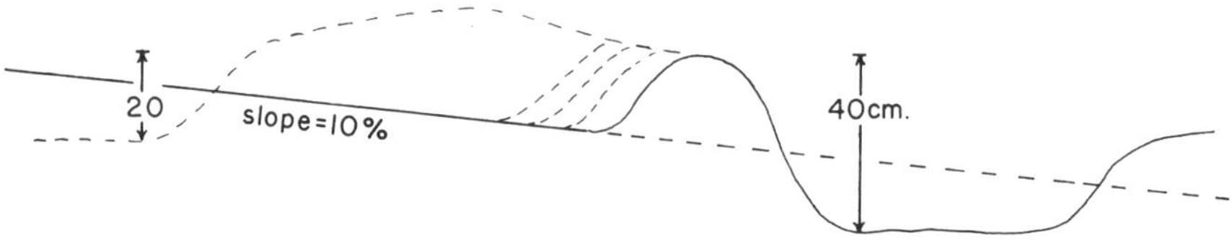

Figure 7-3. *Camellón* construction on hillside.

Figure 7-4. Transition (arrow) from *tablones* to *camellones* on hillside, Totonicapán, Guatemala.

(tablones) that are common in this area (Figures 7-4 and 7-5). The relatively stable *tablones* are less subject to erosion and thus better suited to steeper slopes than are the deeply hoed *camellones.*

Mounded *camellones* are among the most impressive minor field features of Middle America. The volumes of material involved in their construction are also impressive. The cross-sectional area of the *camellón* shown in Figure 7-1 is about 0.375 m^2. At the indicated spacing of one row of mounds every 2 m, a *cuerda* of land contains 220 m (linear) of *camellón.* Thus, if the young farmers' claims are correct, they loosen and shape 82.5 m^3 of material in a working day, strong testimony to the

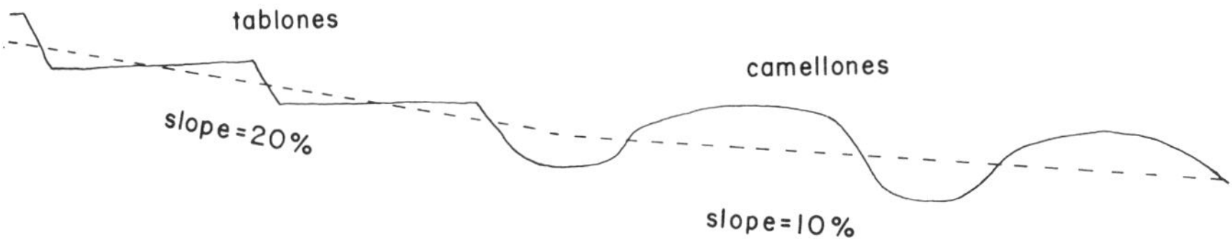

Figure 7-5. Transition from *tablones* to *camellones* as slope gradient decreases.

efficacy of hoe cultivation.[4] At this rate, it would take 22 or 23 man-days to form 1,875 m^3 of soil into a hectare of *camellones.*

At planting time holes are opened with *azadones* on both of the sloping *camellón* sides at about 120-cm intervals (or the length of an *azadón* handle, which is commonly used for spacing). Planting holes are staggered with those on the opposite side to allow sufficient distance between plants while saving valuable bed space. In the Totonicapán region of Guatemala six to seven seeds are placed in each hole, which is then filled and tapped firm with the back of the *azadón* blade. Farmers expect an average of five plants to survive. Using the dimensions and spacings from the example, this would produce a density of about 40,000 plants per hectare. Thus, although *camellones* are relatively large forms with wide plant spacing, plant densities are comparable to those found in other popular systems in Middle America.

Despite the massive reorganization of field surfaces involved in *camellón* construction, further modification is customary. As maize seedlings emerge, farmers form a separate mound *(mul)* around each set of plants (Figure 7-6). Like the maize mounds that will be discussed later, *mules* are customarily built in stages during successive cultivations. Their intended purpose clearly is to protect against lodging (wind throw) because the *camellones* themselves provide well-drained soil and other benefits of raised beds. Because *mules* overlap *camellón* sides, their exact dimensions are difficult to determine. The ones I inspected extended 20 to 30 cm above *camellón* surfaces and were 40 to 45 cm in diameter, for a conservative volume of 0.015 m^3 each for *mules* such as those shown in Figure 7-1.[5] Thus, to the 80 m^3 of soil/*cuerda* (1,875 m^3/

[4]Data from elsewhere support these high production rates. Hoe-formed yam mounds of the Dagomba in Ghana are about 2.5 ft (0.8 m) high, and 4 to 5 ft (1.2 to 1.5 m) in diameter. When the height and radius of a mound are about equal, its shape falls somewhere between a right circular cone and a hemisphere. Applying mound dimensions to both forms and averaging the values, I calculate that Dagomba mounds range from 0.4 to 0.7 m^3 each. "It is said that a Dagomba man normally makes 100 to 200 such mounds in a day." William Allen, *The African Husbandman,* p. 236.

[5]Maize-mound volumes calculated by the procedure described in footnote 4.

Figure 7-6. *Camellones* with *mules* after maize harvest. Stalks are chopped and turned under at next field preparation.

Figure 7-7. Mango seedlings on mounds, Tehuantepec, Mexico.

ha) in *camellones* must be added 5 to 6 m^3 of *mules/cuerda* (120 to 130 m^3/ha) that Guatemalan farmers work and shape each season.

Mounding is not exclusive to annual crops. For example, in Tehuantepec, Mexico, huge mounds a meter or more high elevate young mango trees above poorly drained field surfaces that flood during the rainy season (Figure 7-7). These mounds are not only considerably larger but also more permanent than those used for field crops.

Minor Forms

In the same general region of western Guatemala, ridge-and-bed fields are constructed on moderate to intermediate slopes of 15 to 35 percent. Although vaguely resembling *camellones* in form and planting arrangements, ridges-and-beds[6] are low, sloping, seasonal terraces designed to reduce runoff and erosion and to conserve moisture. Low ridges 15 to 20 cm high and 50 cm wide at the base are constructed parallel to slopes at intervals of about 120 cm. Maize is planted between ridges, in the same staggered formation described for *camellones,* at intervals of 100 to 120 cm (Figure 7-8). After plants achieve some stature, farmers mound individual *matas* (plant groups). However, the ridges remain to combat erosion.

Distinctive checkerboard or waffle patterns of ridges and pits appear in the fields around small towns in the department of Sacatepequez, Guatemala. Maize and other crops are planted on the raised dividers. The completely closed square pits *(cepas)* catch precipitation and encourage percolation in the deep volcanic soils of this region.

To prevent erosion in their maize fields, Pokoman farmers in Santa Cruz Chinautla, Guatemala, prepare flattened round areas (*platos,* literally, dishes) with hoes, then plant maize with wooden sticks. Several varieties of squash also serve as ground stabilizers and erosion inhibitors in the *milpas* (maize fields).[7]

From central to western highland Guatemala narrow, handmade, temporary ridge terraces are formed parallel to contours on steep (up to 60 percent) slopes (Figures 7-9, 7-10, and 7-11). Spacing of about 1.5 m appears to be almost independent of gradient. Small checks every 1 to 2 m prevent water from running in the furrows. Despite the massive reorganization of field surfaces, ridges and furrows must be reformed each season because normal field preparations destroy them.

[6]Because I found no consistently used term for these features, I use the descriptive *ridge-and-bed.* Local usage includes *camellones* for the ridges and borders of these plots.

[7]Ruben E. Reina, *The Law of the Saints,* p. 47.

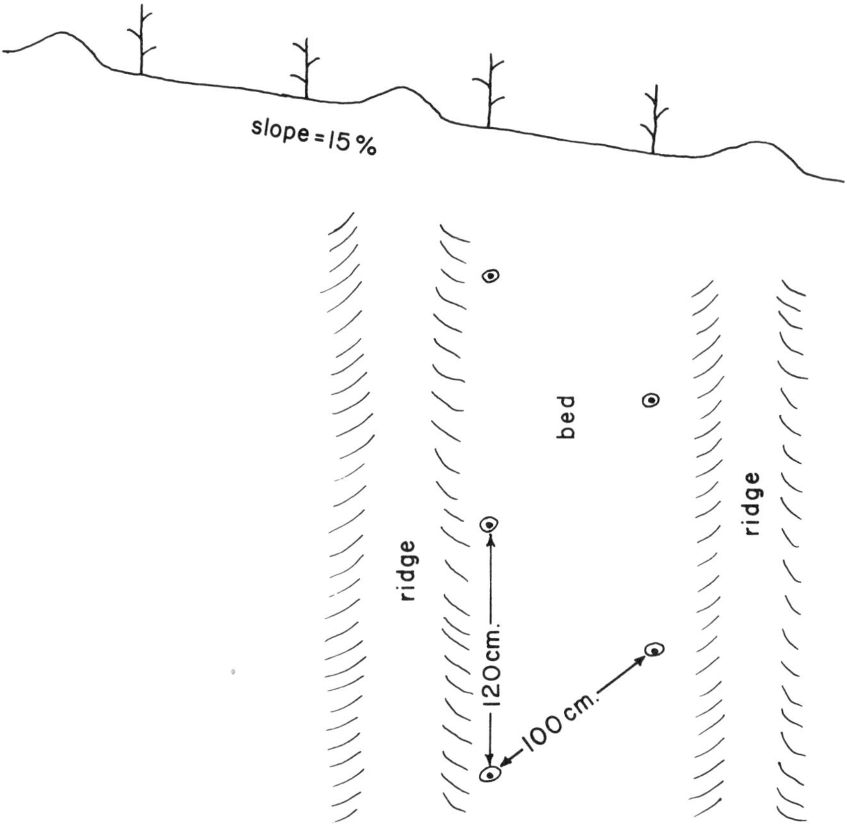

Figure 7-8. Ridge-and-bed structures.

Deeply furrowed fields on the lower slopes of La Malinche, central Mexico, resemble the ridge terraces of western Guatemala (Figure 7-12). But slopes are gentle and furrows are formed with team-drawn plows rather than hoes.

Maize Mounds

Microtopographic features associated with maize cultivation account for the bulk of field-surface modifications in Middle America, in both number and extent. Of these, the most common are the hills or mounds of earth heaped up around the bases of maize plants. Maize mounds are found all over the region and go by a variety of general names such as *aterradura, matilana, calzada, montón,* and *mogote,* together with appropriate verb forms: *aterrar* or *matilanear* (to draw up

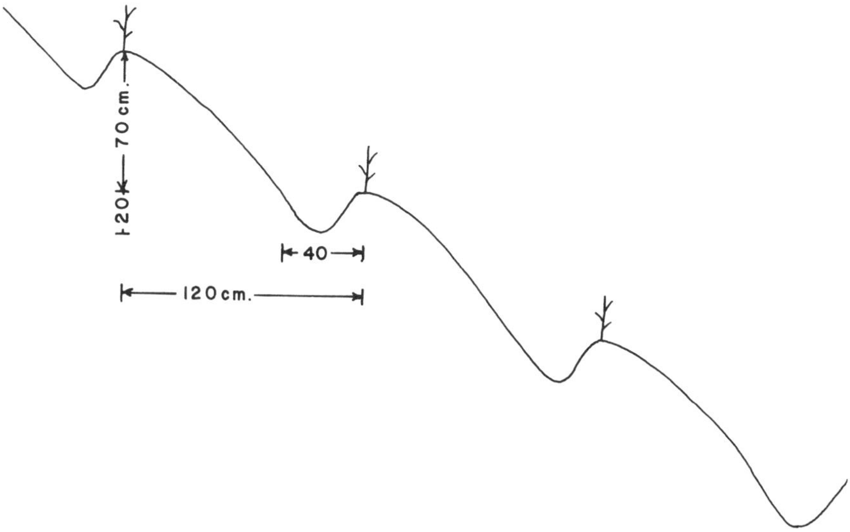

Figure 7-9. Temporary ridge terraces.

Figure 7-10. Detail of ridge-terraced field, Sololá, Guatemala.

Figure 7-11. Steeply pitched ridge-terraced field, Sololá, Guatemala.

Figure 7-12. Deep plow ridges and furrows in maize field, Puebla, Mexico.

Figure 7-13. Maize mounds, central Mexico.

earth), *calzar* (to put on shoes), and *echar montones* or *mogotes* (to throw up mounds or hummocks). In addition, the ubiquitous mounds also are identified by a great number of local names (for example, *campana,* referring to their bell shape) and in the many indigenous languages (for example, *mul* in Quiché, a Maya dialect).

The practice is ancient. Hilling is mentioned frequently in the contact-period literature and is pictured in the codices:

> And when [the seed] had sprouted and formed shoots, [the gardener] hoed the ground; he built up mounds; he broke the clods of earth, and went stirring up the dust.[8]

Maize mounds (Figure 7-13) are developed during normal cultivation cycles. Customarily mound formation is a two- or three-step process: low mounds are formed around young plants during the first or second cultivation, then raised to final form in the second or third cultivation. In team-worked fields maize is usually planted in furrows that are converted into ridges by the first cultivation. These are broken down into mounds around each *mata* with hoes in the second and third

[8]Bernardino de Sahagún, *General History of the Things of New Spain,* Pt. 4, p. 128.

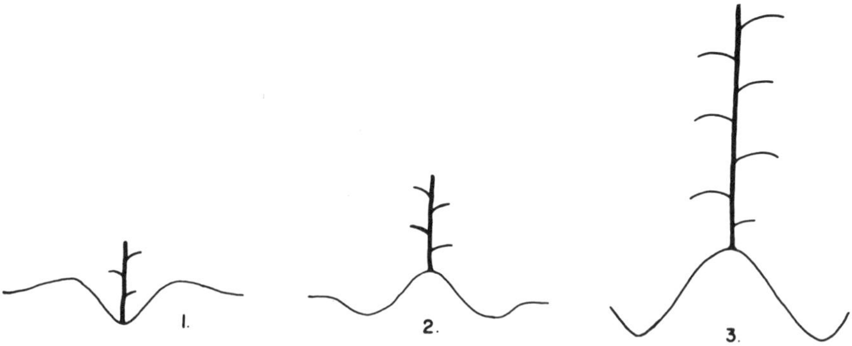

Figure 7-14. Three stages of maize mound formation: (a) maize planted in furrows; (b) furrows converted to ridges in first cultivation; (c) ridges broken into mounds around individual plant groups.

workings *(labores)* (Figure 7-14).[9] In all cases mound formation is associated with maize cultivation *(labores de la milpa)* rather than field preparation *(labores de la tierra).*

Although maize mounds are broken down and reformed each growing season, under certain circumstances they may become fossilized into relatively permanent features in the landscape. For example, encroaching grass in old fields abandoned to pasture can preserve mounds for years and possibly centuries (Figure 7-15).[10]

Size of maize mounds varies greatly. Field measurements indicate a general range of 50 to 80 cm in diameter and 25 to 45 cm in height (compacted, not newly formed). Because large mounds attract more attention, there may have been some bias during the survey against fields with smaller mounds. But even if a true average falls toward the lower end of the range, there must be material on the order of 0.03 to 0.05 m^3 in each mound.

Despite the fact that maize is the single most important crop in Middle America, little information is available on traditional plant spacing or field densities. Based on field counts, I derive a representative

[9]Apparently a similar process was followed in pre-Spanish cultivation: "At this time it [maize] is hilled, the hollow is filled in, the crown is covered, the earth is well heaped up." Sahagún, *General History,* Pt. 12, p. 283. In some modern systems where mounding has long been abandoned farmers still refer to individual plants or plant groups as "hills of corn." The Spanish *mata* conveniently describes plant clusters without reference to planting system and thus avoids this confusion. For clarity the terms *mound* and *hill* are used here only when soil is actually heaped up around the base of plants.

[10]Paul Weatherwax, *Indian Corn in Old America,* pp. 70–72.

Figure 7-15. Grass-covered fossil maize mounds, Quezaltenango, Guatemala.

figure of 11,000 to 12,000 mounds/ha.[11] Multiplying by the conservative estimates of volume, we find that mounding a hectare of maize requires transferring and shaping 330 to 600 m^3 of soil, no small task considering that it is accomplished in large part by hand.

Labor costs for mounding are therefore significant. In central Mexico a fair or normal day's work *(tarea, jornal)* expected of hired laborers *(peones)* is 800 mounds or *matas* in fields already ridged with plows (that is, step (c) in Figure 7-14) and planted only to maize.[12] At

[11]Mound formation and spacing are not unrelated. Several informants explained that the common distance of 80 to 100 cm between *matas* provided the minimum surface area from which soil could be taken to form hills without digging into root zones. In traditional Middle America several maize seeds are customarily planted in the same hole with the expectation that an average number will germinate. Thus, plant density is the number of surviving plants per hill times number of hills per hectare. In the example, if three or four plants survived in each hill, plant density would be about 35,000 to 40,000 plants/ha, a figure close to optimum densities determined in government experiments. Felix Ramírez P. and R. J. Laird, *Densidad Optima de Plantas de Maíz para los Valles de México y Toluca.*

[12]John K. Hatch (*The Corn Farmers of Motupe,* pp. 108-9) reports that in Peru a day's work of hilling with a shovel *(aporque a palana)* is about 2,500 mounds, a high quota compared with those of central Mexico. He describes mounding as "the most physically strenuous task of the corn crop cycle."

this rate, a hectare of 11,000 to 12,000 *matas* would take 14 to 15 man-days to complete. Mound formation in fields interplanted with beans or other crops requires more care, and quotas drop to 700 mounds or fewer. Adjustments are also made if soil or moisture conditions impede steady work. A somewhat different, two-stage process in Guatemala produced confirming estimates of labor costs. Without prior plow ridging, low maize mounds are formed with broad hoes about 4.5 months after planting, then are enlarged during a later cultivation. One man can mound a *cuerda* of 440 m^2 at either stage in one-half to three-quarters of a working day, or a hectare in 12 to 17 days. In both Mexico and Guatemala mounding customarily is done in small groups—the farm owner *(dueño),* his sons or other male relatives, and perhaps a hired or exchange laborer or two—so that a hectare is completed in a few days.

With these data and going rates for field labor, the cost of mounding maize could be determined fairly easily. But most work is done by farmer/owners and their immediate families, and labor costs are reckoned only when actually paid as wages to hired workers. Furthermore, many crops, especially maize, are destined for home consumption, making the relationship between production costs and returns unclear. Finally, industrious farmers often have a habit of toiling on at tasks that could be done by inexpensive hired labor. Many times I have watched relatively affluent farmers, ignoring opportunities for well-earned leisure, spend additional hours at laborious chores, such as mounding maize, that could have been delegated or even neglected.

But if the costs of maize mounds are a problem to calculate, their benefits are well-nigh impossible to ascertain. Not only are there almost no data on the effects of mounding, but some even question whether mounding has any useful effects at all! Traditional farmers themselves most commonly mention protection from lodging as the primary reason for mounds. The highlands of Middle America are subject to strong winds during the maize growing season. Sharp gusts associated with convective storms are especially feared because they can flatten a *milpa* in minutes. Even farmers who do not mound all the *matas* in their fields may shore up the outer two or three rows and hope that these will shield the interior.

From this perspective, the costs of mounding maize are crop-insurance premiums to be weighed against the probability of loss. Wind damage, especially from thunderstorms, tends to be localized,[13] and farmers may be willing to gamble, especially if mounding costs increase. This appears to be the case in Tlaxcala, where industrial labor demands have driven wages for field labor well above the legal minimum. Many farmers are abandoning the practice of mounding except in those plots

[13]Don Bostwick, *Studies in Yield Variability.*

where they can do the work themselves. At 14 to 15 man-days/ha, the value of maize mounds as insurance against wind damage has become marginal.

Other reasons for not mounding apply in specific instances. Hillside fields with poor soils and stunted plants are not treated because "the short plants are not affected by winds" or "the crop is not worth the work of mounding" (a variation of the cost/benefit approach). Some farmers hold that the effort is wasted (that mounds are ineffective) in very sandy soils. A few varieties of maize are believed to have such strong root systems that mounding is unnecessary. Short-cycle, repeated crops, such as three-month maize grown for green roasting ears *(elotes)* usually are not mounded because of the amount of labor that would be required to mound twice or even three times each year. Furthermore, *elotes* are a cash crop, and traditional farmers seem more willing to take risks with market crops than with subsistence maize.

In addition to protecting against wind throw, maize mounds may have other functions, most of them associated with improving soil conditions. They are believed to promote tillering, to improve soil drainage and aeration, to decrease soil moisture losses from evaporation, and to suppress weed growth or facilitate weed control.

But perhaps the most astounding feature of this ancient and widespread practice is that some authorities, and some traditional farmers too, do not believe that mounding is effective against lodging or that it is beneficial in any other way. Thus, more than 100 years ago Enfield was openly disdainful of the practice:

> The useless habit of piling up the earth in cone-shaped hills around the stalks of corn, which was at one time almost universal, is now generally disapproved and wisely abandoned. It was formerly supposed to aid the stalk in resisting the effect of severe gales, but experience has proved this to be a mistaken notion. There was also an imagined advantage in drawing up the earth around the roots; but here again experience has developed the sounder philosophy of allowing the roots to find the earth, as they require it, by their own spontaneous movement. This they will be sure to do; and they will find the manure also, provided both manure and soil have been sufficiently pulverized and blended.[14]

I have talked with many farmers who would agree with Enfield that mounding is an antique practice of little or no value and certainly not worth the considerable time and energy it requires.

Some authorities go even further and claim that mounding is detrimental. These arguments focus mostly on the likelihood of exposing or cutting shallow roots when scraping up soil to form mounds. But, in addition, mounding may permit soil bacteria to attack

[14]Edward Enfield, *Indian Corn,* p. 140.

maize stalks 30 to 40 cm above the base, where protective outer layers are thin, thus weakening the stalks and making them more susceptible to wind throw than they would have been without mounding.[15]

This situation is most confusing. Traditional wisdom, supported by centuries of practice and uncounted man-hours of effort, argues for maize mounding. Yet contemporary opinion, including that of many traditional farmers, disparages the technique. We are left with the appealing but unsatisfying explanation offered by Weatherwax:

> Why does the Indian persistently cling to this way of doing things? And does it [mounding] have any merit? The Indian's answers to such questions are vague. It prevents the water from standing around the plants; or it supports the plants and prevents wind damage. But his clinching answer usually is that he gets more corn when he does the thing this way, and that this is the proper way to grow corn anyhow.[16]

Irrigated Fields and Seed Beds

Many of the techniques already discussed for rainfall-irrigated *(temporal)* plots, such as mounding or ridging, work equally well in irrigated plots. But the need to control irrigation water gives rise to another group of minor features. Of these, perhaps the most common are plow-formed ridges and furrows associated with furrow irrigation, and the checks or borders around flood-irrigated fields. Because surface irrigation water often flows with sufficient force to erode, checks or borders are usually permanent features of substantial size. But where water application is carefully controlled, temporary and relatively fragile microforms suffice. Such minor features are found, for example, in manual-irrigation regions where water is hand carried or splashed from wells or canals.

Container-irrigated fields are usually divided into a series of narrow plots *(canoas, melgas, tablones)* no more than 2 m wide. These are subdivided into tiny compartments *(cajetes),* some only 50 cm wide and long (area = 0.25 m^2), each containing but a few small plants such as onions or garlic (Figure 7-16). Because water is applied carefully by hand, retaining borders around plots and checks need be only a few centimeters wide and high, just enough to contain a thin layer of water during any one irrigation. Borders around plots are carefully formed with hoes or shovels. The tiny edges around individual checks are shaped and patted firm with bare hands.

[15]John E. Hibbits, personal communication.
[16]Weatherwax, *Indian Corn,* p. 70.

Figure 7-16. Carefully formed *tablones* broken into *cajetes* to contain hand-delivered irrigation water, Oaxaca, Mexico.

For large plants, individual ridged planting beds or pits *(cepas)* are formed. The hollow mounds are designed to hold up to three tomato plants. As soon as seedlings are transplanted from seed beds, *cepas* are formed around them to confine irrigation water. The system effectively conserves the water arduously extracted by hand from shallow wells; wetted surfaces within the *cepas* constitute only about 20 percent of total field area. Planting in pits is not limited to annual crops. Seedling trees, such as avocado, initially are set in *cepas* at proper rooting depths. The pits are not bordered but are left open to collect debris and precipitation, and to facilitate fertilizing and bucket or hose irrigation; they gradually fill in by deposition as the trees age.

Sprinkle- and splash-irrigated fields require even less in the way of retaining structures because water amounts and rates are completely controlled. Thus, splash-irrigated plots have no borders at all or, at most, are bounded by small ridges.

Seed beds *(almácigas, sementeras)* are used widely in traditional Middle American agriculture, especially where space or growing seasons are limiting factors. Seed beds are small, no more than a few square meters, and are built with consummate care. Initial preparation is with team-drawn plows or by hand with hoes and shovels. A variety of materials including sand, plant residues, manures, leaves, and chemical fertilizers are mixed with seed-bed soils to improve tilth and fertility. Soils are carefully raked, sifted, and examined for sticks, rocks, clods, or other impediments to germination and seedling development. Beds often are raised slightly to improve drainage and are covered with surface mulches to shade the surface or reduce heat and moisture losses from the soil. Like other hand-tended planting surfaces, seed beds need only low borders to retain irrigation water.

Summary

This brief list includes only a few of the minor field forms in Middle America. Over the centuries traditional farmers have developed an enormous variety of mounds, ridges, pits, furrows, and beds, each of which is a response to particular crop, climate, soil, and slope conditions, and all of which presumably improve crop plant environments. The term *presumably* is chosen deliberately. Clearly such forms as water-containing ridges and pits control distribution and loss of scarce water and are of great value. The effectiveness of other minor field forms is not so apparent. Although calculating labor costs and other inputs is relatively easy, determining results requires precise field measurements and yield accounting—techniques not yet used in traditional-agriculture research. Without such data there is no way to evaluate these centuries-old practices.

Yet the value of such analyses would be enormous. If the results of minor field forms do not justify costs, then millions of man-days could be saved annually by abandoning them.[17] However, if microtopographic adjustments to field surfaces result in higher production or lower losses, then their continued and increased use should be promoted.

[17]For example, Manuel Gollás ("Mano de Obra Excedente y Eficiencia Económica en el Sector Tradicional de una Economía Dual") argues that too much family labor is used in traditional farm activities.

8

Water Management with Intermittent Sources

General Introduction

A clear distinction is usually drawn between managed and unmanaged water systems. Unmanaged systems depend on natural water deliveries and removals, whereas managed systems involve significant human intervention. But, as is true of so many basic concepts, close scrutiny reveals blurred distinctions. In the world of traditional agriculture, water management ranges from complicated irrigation systems with elaborate dam and canal networks to practices so simple they seem hardly like deliberate management at all.

The key phrase here is deliberate management. Exploitation of naturally favorable environments, such as moist windward slopes or naturally irrigated alluvial fans, are excluded from the discussion as are practices that unintentionally alter the distribution of water, such as devegetating catchment basins.

Three factors are crucial to understanding agricultural water management in traditional societies. The first is the mass of water involved. For reasons still unclear plants use much more water than they seemingly need for biological processes.[1] In addition, much of the water entering farm ecosystems is never used by plants at all but escapes

[1]Nyle C. Brady, *The Nature and Properties of Soils,* pp. 207–10.

as surface runoff, deep percolation, or evaporation. Consequently, the total amount of water that passes through a farmer's field during a growing season is enormous, exceeding, in many cases, the mass of all other resources, including the tilled soil layers themselves.

The second factor is the level of energy in traditional agriculture. By definition these are low-energy systems that use little or no fossil fuel. Most traditional water management in Middle America uses human labor, either for moving earth to construct or repair dams, canals, and field borders or for irrigating crops by hand.

The two factors discussed so far, volumes of water required in agriculture and lack of power for lifting or transferring water, would be strong deterrents to water management in low-energy farming systems. The third (and saving) factor is, of course, the water itself. Because it is a liquid, water is infinitely divisible and can be collected, stored, transferred, and applied in any quantity. But most important, the slightest gradient allows water to move on its own. All that is required is to channel and carefully adjust the grade. There are few situations where water is conveniently located on a grade above fields. More often it is necessary to build small dams to raise stream levels and to construct canals to carry water to the fields. The challenge is to direct water so that it will move on its own but not flow so much as to cause erosion.

Water not only runs downhill but it can also move itself upward by capillary action. Middle American farmers have excelled in developing subirrigated farming systems in areas where high ground-water tables supply moisture to crop root zones by capillary lifting.

But where neither gravity nor capillary transfers can be used, water must be lifted by some other means. Middle American farmers have been reluctant adopters of wind- or water-driven mills to lift and transfer water and of the animal-powered devices that are so common in North African and Asian agriculture. The alternative of lifting by using human power is so costly that only small areas can be watered. Nevertheless, manual irrigation systems play a prominent role in Middle American agriculture.

The following review is divided into systems with intermittent supplies of water (discussed in this chapter) and those with continuous supplies (discussed in Chapter 9). Subsurface water supplies are treated separately in Chapter 10 because they have different management characteristics and also because of their distinctive role in past and present Middle American agriculture.

There is no ranking according to level of sophistication of water management. The best solution to a given water management situation is the one that gets the job done at least cost (effort). Ordering the various practices by complexity might suggest an evolutionary or developmental sequence that could not be supported.

Rainwater Management

By far the most extensive farming systems are those that depend solely on precipitation. There is little that is possible in the way of management of atmospheric moisture, and Middle American farmers can only adjust crop types and combinations and farm schedules to the seasons and uncertainties of the climate.

On impact with the ground surface, however, precipitation is subject to a variety of traditional management practices. Shelters, arbors, and mulches protect tender seedlings and reduce splash erosion and surface compaction by absorbing the impact of falling rain or hail. The effects of various tillage practices are subject to some controversy.[2] However, traditional farmers are convinced that plowing and cultivating "open the soil," that is, they are effective means for increasing the rate and amount of rainwater absorption. Furrows, ridges, and terraces increase infiltration and retard runoff and erosion.

Runoff Management

Horizontal flows of water on the surface are much more amenable to management than are the uncontrollable and unpredictable patterns of precipitation. Even so, farmers in most rain-watered systems are content with the natural configuration and drainage characteristics of their agricultural lands. But where rains are scant or slopes are steep, altering gradients to impede surface runoff or diverting flows to catchments may be warranted.

One of the simplest methods for retarding surface runoff is to simply lay logs or other debris across the slope of fields. This practice is commonly used in shifting cultivators' plots where materials are conveniently at hand. Often such alignments are not apparent in the tangle of burned and unburned debris. This ancient practice continues today in areas where the dangers of sheet and rill erosion warrant the effort.

From using rows of debris it is only a short step to installing lines of rocks to check surface runoff. The extra effort of moving and positioning rocks is offset by the permanence of the structure. Even in fields worked for only a few years before fallowing, it probably requires less total effort to install rock check strips than to replace burned and rotted logs and brush rows, especially if the rocks already are in the

[2]Brady, *Nature and Properties of Soils,* pp. 66–68.

field. Neither logs nor lines of rock offer serious impediments to hand cultivation.

Log, brush, or rock check strips retard surface runoff and increase infiltration. This results in the settling out of suspended sediments to form deposits that gradually build up behind check strips to form crude terraces. Apparently, the rows of rocks and suggestions of terraces in long-abandoned Maya fields of eastern Belize are remnants of this ancient practice.[3] Building stone field borders to control runoff may have been one of the first management techniques developed in the Tehuacán Valley, where such borders have been dated to A.D. 300.[4]

Log or rock check strips are simple to install but tend to be leaky and easily displaced. Retardation of runoff is improved by continuous earth or earth-and-rock embankments constructed across the slopes of rain-fed fields. Spacing and elegance depend largely on slope conditions and on the industriousness of the farmer. Surface flows are limited to the distance between embankments if embankments are properly constructed. Spontaneous or planted vegetation reinforces the structures, especially against the erosive effects of ponded water spilling over the top. Over the years soil transfers associated with plowing and sheet erosion build up behind the embankments to create semiterraces. Earthen embankments also increase crop diversification; these separate spaces, within but distinct from fields, are often planted to perennial or exotic crops.

But as management increases, so too does the possibility of breakdown of structures and the need for careful engineering and maintenance. Casual log or rock check strips only imperfectly control runoff, and debris transfers continue in attenuated form. Although erosion will increase if check strips are disrupted, it will not usually increase radically. But behind continuous embankments surface flows are effectively stopped and water is ponded (an incipient storage phase). Lateral drainage may be necessary, which requires careful gradient adjustments and provision for off-flows. Ponded water on a slope represents potential energy that, if released by spilling over embankments or flowing through ruptured banks, will be concentrated on a small portion of the field and cause gullying.

Erosion hazards are reduced and infiltration is increased by adding special water-storage structures. For example, over much of Middle America farmers excavate collection ditches *(zanjas)* just down slope from embankments and along field edges perpendicular to the slopes. Those down slope collect seepage and sheet wash; those along field edges catch lateral runoff that otherwise would pour down paths and

[3]C. Earle Smith, Jr., "Agriculture, Tehuacán Valley."

[4]Richard B. Woodbury, "Prehistoric Water Management Systems in the Tehuacán Valley, Mexico."

droveways. In both instances silt deposits must be removed periodically, usually to be spread as top dressing on adjacent fields.

The total storage capacity of a *zanja* system is substantial. Using the figures for the *zanja* operation in western Tlaxcala that is described in the section on sloping terraces in Chapter 6, we can estimate that a hectare of land would have a water-storage capacity of almost 400 m^3. That is, in the drains alone, not including ponding behind raised embankments, the equivalent of 40 cm of precipitation could theoretically be stored. In actuality most *zanjas* are partly filled with debris from deposition and bank collapse. Nevertheless, they are an important and effective method for managing rain water on sloping fields.

Rows of logs or rocks or raised embankments reduce erosion and increase infiltration by checking surface runoff. The same effect is achieved by reducing overall gradients in fields. Casual down-slope debris transfers have already been mentioned. Over the years material eroded from higher portions of fields is deposited behind rows and embankments to create sloping terraces or semiterraces. In some systems farmers accelerate the process by consistently turning furrows downhill so that soil is gradually worked down behind the retaining embankments. Thus gradient reduction for water management is assisted both by gravity-driven sheet wash and by gravity-assisted plowing.

Check dams built in intermittent streamways also reduce gradients. Dams or rock, earth, or brush reduce channel runoff speed and induce deposition. The flat alluvial fields that subsequently build up behind the dams are flood irrigated or subirrigated by flows in the streamway following storms. Check dams thus offer a striking example of how traditional farmers manage natural flows and forces to achieve desired ends. In this case the usually negative forces of erosion and runoff are manipulated first to create fertile fields, then to irrigate them and renew their fertility by flooding and silting.

Field gradients can also be reduced by direct transfers of soil. Field leveling is not uncommon, and flat or bench terraces are found in many parts of Middle America. But such projects are expensive in traditional systems and are undertaken only where land or crop values are high.

Runoff Irrigation

Systems that concentrate overland or stream runoff to flood cultivated fields are widespread in Middle America. They range in size from individual fields watered by runoff from adjacent slopes to basin-size systems in which natural drainage is rechanneled into complex distribution networks.

In its simplest forms runoff irrigation is not so much management as exploitation of naturally favorable environments. As Bryan notes, "the usually clear distinction between irrigated and non-irrigated farm land breaks down in the consideration of marginal lands irrigated by flood water."[5] Essentially, precipitation falling on cultivated fields is augmented by runoff from uncultivated slopes above. Bryan goes on to describe some of the characteristics of fields on which this type of irrigation can be used:

> The selection of a field involves an intimate knowledge of local conditions. The field must be flooded, but the sheet of water must not attain such velocity as to wash out the crop nor carry such a load of detritus as to bury the growing plants. Such conditions require a nice balance of forces that occur only under special conditions. Shrewd observation and good judgement are necessary in the selection of fields.[6]

Sites chosen fall into two main types: below slopes with poor infiltration (rocky, denuded) and on alluvial fans or valley floors where runoff is delivered by natural drainage channels. Such incidental runoff irrigation is widespread throughout Middle America, especially in any areas where more complex systems have not been established.[7]

Advanced runoff-irrigation systems may include collection and transfer structures that give farmers considerable control over flood waters. Thus, on alluvial fans low diversion dams are sometimes built to spread water from relatively easily managed intermittent streams. Short ditches may direct water to fields, which themselves may be checked or bordered. Although only slightly more complex than uncontrolled flood-water irrigation, these systems allow farmers to build or break diversion dams and to open or close canals and thus to decide whether to irrigate. Two such systems in the Tehuacán Valley are described in detail in Chapter 5, in the section on controlled flows.

A similar though somewhat smaller system is operated by Zapotec farmers in the town of Díaz Ordaz (also Santo Domingo del Valle), in the Oaxaca Valley of Mexico.[8] Flood waters from eroded swidden (slash-and-burn) plots of an upstream village are checked by six diversion dams across an intermittent streamway. Dams are about 2 m high and 20 m long, constructed of logs interlaced with saplings and plugged with rock and earth fill. The dams channel water through a canal head and down main canals divided into two to twelve subunits, or *tramos*.

[5]Kirk Bryan, "Flood-Water Farming," p. 444.

[6]Bryan, "Flood-Water Farming," p. 445.

[7]Gary Paul Nabhan, "The Ecology of Floodwater Farming in Arid Southwestern North America"; Carl O. Sauer and Donald Brand, *Prehistoric Settlements of Sonora with Reference to Cerros de Trincheras.*

[8]Theodore L. Downing, "Irrigation and Moisture-Sensitive Periods."

Tramos are the smallest administrative unit. Each *tramo* has thirteen to forty-one members and three officers: president, secretary, and treasurer. Each parcel owner takes office for one year in rotation. *Tramo* administrators mediate disputes, assign obligatory maintenance work, collect a small tax, and distribute water allocated to the subsections by the *sindico,* or local authority that coordinates the more than seventy *tramos* of the area. Depending on supply, the *sindico* apportions water according to either riparian rights, where proximity of land to water source determines priority, or prior appropriation, where historical precedence rules.

Summary

Traditional water management without storage begins even before the rain reaches the surface and continues with a variety of simple to complex practices to retard overland flows and divert runoff into irrigation and silting channels. Even the simple techniques test traditional ingenuity. The substantial quantities of water that must be controlled, and the ever-present danger of erosion should controls fail, pose stiff challenges to farmers who must confront these powerful forces with the limited supplies of energy and materials that are available locally.

Although runoff irrigation is ancient,[9] as with so many traditional practices it survives and thrives in contemporary times.[10] One reason for the continuing vitality of these systems is that they require relatively little in the way of construction and maintenance. In its simplest forms irrigating with flood water almost escapes being classified as management and becomes one of those practices associated with exploitation of naturally favorable environments. Even such highly managed works as the *presa/zanja/pantle* complex of Tehuacán do not involve the labor investments and operating expenses of comparable irrigation systems that include storage.

[9]Runoff irrigation may have preceded farming. Julian H. Steward ("Irrigation Without Agriculture") describes a system in the western United States consisting of a dam and two ditches, and irrigated fields that were neither planted nor cultivated. Instead, flood waters were diverted to irrigate wild plants that were subsequently gathered.

[10]For example, Kirkby (*The Use of Land and Water Resources,* p. 36) notes that "floodwater farming is the most widespread type of water use practiced in the [Oaxaca] Valley and covers the greatest range of physical environment, scale, complexity, and frequency of use."

All runoff systems, however, remain tied to immediate weather events, which severely limit the decision-making options of farmers. The next step is to develop storage facilities and gain control over timing of water delivered to fields. But this logical extension not only demands more time and effort but also involves allocation procedures that have profound social and political implications.

9

Water Management of Surface Systems

Although some runoff-irrigation systems are quite complex, lack of storage or a steady flow of water severely limits the range of management choices. Even a small dam and pond on an *arroyo* provides a degree of control over the supply of water and many traditional irrigation systems have some form of storage. But traditional dam and canal technology is limited, and most of these structures are small. Although these structures may even out flows and save water that otherwise might be lost, few are large enough to store enough water to last for a season or even for a prolonged dry spell. Also, most systems depend on supplies from springs, streams, or water tables. A dependable supply allows farmers to decide when and how much to irrigate.

In this chapter I shall review methods that apply to gravity transfers as well as several systems where water is below field level and must be lifted.

Sources of Water

Runoff with Storage

Among the common forms of surface storage are simple structures to impound and discharge surface runoff in channels of intermittent streams.[1] Usually these structures consist of small earthen dams, sometimes reinforced with rocks or branches. However, because channels are seldom enlarged, the amount of water stored is limited by the natural configuration of the streamway. Most of the thousands of runoff reservoirs in Middle America simply impound water for stock. Those that are used for irrigation have log or concrete frames at the bases of dams and easily lifted wooden-board or sheet-metal gates that permit the regulated discharge of water. Runoff storage systems are usually small and discharge only into a single canal network.

Yet even these simple arrangements present traditional water managers with an array of challenges. Dam design and construction must be adequate to resist often turbulent runoff. Of particular importance are spillways, which permit discharge of superfluous water that would otherwise spill over or damage dam structures. Traditional farmers early on discovered the reinforcing virtues of rock and grass on spillway surfaces. Farmers also face high evaporation losses in reservoirs. In addition, reservoirs fill with deposits, and silt removal is a difficult task for farmers without powered equipment. Like modern water managers, traditional farmers may find it more expedient to build larger dams downstream than to excavate old silt deposits.[2]

Springs

Among the more manageable water sources in traditional farming are springs, which appear where ground surfaces intercept water-bearing strata. If springs broach well up slope, so much the better, for then water can be transferred by gravity to fields below. Spring flows often are small and many are channeled directly into canal systems. Such is the case in several areas of western highland Guatemala, where canal grids bring water only to field edges; farmers then scoop or splash it onto plot surfaces.

In other areas, small flows may fluctuate or be inadequate for efficient irrigation. Reservoirs are then needed to regulate flows and to accumulate water that can be released quickly to irrigate an entire field in a short time. Reservoirs also save water that might otherwise be lost during inconvenient irrigation times, such as at night. Simple dams and

[1]T. M. Boers and J. Ben-Asher ("A Review of Rainwater Harvesting") review recent literature on runoff inducement, collection, storage, and farming. Future prospects are discussed by Boers and Ben-Asher and by the National Academy of Sciences, *More Water for Arid Lands,* pp. 9ff.

[2]In one of the oldest irrigation systems in Middle America successive generations of farmers simply built higher and higher dams to compensate for storage capacity reduced by siltation. Richard S. MacNeish, "Ancient Mesoamerican Civilization."

reservoirs adequate for most spring flows are well within the limits of traditional technology. For these reasons storage structures are common features on traditional spring-fed plots.

The apparent point source of spring water also facilitates administration. In most cases, villages on whose lands springs appear control the flows and use most of the water. Unless special sharing relationships have been established, downstream villages have only subsequent rights. Water is administered by village committees and supervised on a day-to-day basis by elected or appointed canal or water tenders *(aquadors)*, who receive requests for water from individual farmers and notify them of their turns.

Larger systems are more complex. When spring water is used by more than one village, intervillage coordination and conflict-resolving institutions are necessary.[3] Quantities and timing can make the difference between success or failure in subsistence or commercial operations, and it is not surprising that water disputes and thefts are the source of bitter village feuds. A management system associated with a large spring *(manantial)* in Mexico will illustrate the complications that develop when flows are substantial and are of critical importance to agriculture.

Although the Tehuacán Valley of western Mexico enjoys year-round growing temperatures, water is scarce. Summer rains are adequate for only a single crop and even that is at risk because the amount of precipitation varies. But the valley is blessed with an abundance of springs, and irrigation has been important for more than 2,000 years.[4] The modern agricultural economy is based on irrigated, multiple-crop production of green maize *(elotes)*, garlic, tomatoes, and other specialty crops that are sold at lucrative off-season prices in Mexico City and other urban markets.

One of the most abundant springs in the valley is Texcali, or La Taza,[5] which rises about 15 km southeast of the city of Tehuacán and goes past the villages of San Gabriel Chilac and San Francisco Altepexi. In 1974 estimated flow from this spring was 488 liters per second (lps),[6]

[3]René Millon, Clara Hall, and May Díaz, "Conflict in the Modern Teotihuacán Irrigation System." A number of studies have used the Middle American experience to focus on the organizing qualities of irrigation—for example, Robert C. Hunt and Eva Hunt, "Canal Irrigation and Local Social Organization"; Karl A. Wittfogel, "The Hydraulic Approach to Pre-Spanish Mesoamerica."

[4]C. Earle Smith, Jr., "Agriculture, Tehuacán Valley."

[5]Most springs in Mexico have Spanish or Christian names (for example, La Taza, the basin), indigenous names (for example, Texcali, house of stone, from the shape of the outlet), or both.

[6]Data from the files of the Secretaría de Agricultura y Recursos Hidráulicos, Tehuacán. The flow of La Taza was earlier reported at 522 lps. Luis Blásquez L., "Hidrogeología de una Parte de Tehuacán, Edo. de Puebla," p. 38. An even

which was divided equally between Chilac and Altepexi.[7] Other springs and *galerías* (horizontal wells) feed into the irrigation systems of Chilac and neighboring towns.

Chilac's share of Texcali water flows into a small oblong basin from which several canals emanate (Figure 9-1). This second division originally allowed water to be diverted to the four *barrios* (quarters, subvillages): Tlacoatl, Tlaconahuac, Ecatzingo, and later Tepetiopa. Four main distribution canals still exist, although, as León Marquez notes, the contemporary *barrios* are not separated in any meaningful way.[8] The basin and the canals that flow out of it form a collection system for the waters of Texcali and for the lesser springs and *galerías.*

The waters continue through a complex canal network to the several *barrios* and on to individual fields. Water from several sources and destined for several users can be combined into a single canal. Accounting is done by *ameyales,*[9] or measures of water. Twigs stuck into channel sides show the depths belonging to each owner *(dueño)* and are not to be exceeded when taking water from canals. The basin and canal arrangement permits rapid, precise transfers of spring and *galería* water in whatever quantities are needed.

There is no man-made storage facility in the system. The generous Texcali spring provides a year-round supply of water and the generally dry climate creates a nearly continuous need for irrigation. Adjustments for relative abundance and scarcity of water during the wet and dry seasons are made through a unique market and price mechanism.

Theoretically, there is no private ownership of water in Mexico.[10] But for centuries the resolute citizens of Chilac have fought to retain control of what they perceive as their water, and the independent spirit

---

earlier report of 225 lps for Texcali probably refers to only Chilac's share. Borja León Marquez, "Breve Estudio de las Condiciones Sociológicas e Higiénicas del Municipio de Sn. Gabriel Chilac, Tehuacán, Pue.," p. 57.

[7]Some present-day patterns of water use and allocation date from the early Spanish period. For example, the original rights to Texcali water given to Chilac in 1543 were later modified (abrogated?) so that several communities received allotments; a formal judgment in Mexico City in 1641 divided the water between Chilac and Altepexi. Gorgonio Gil Huerta, "History of the Foundation of the Town of San Gabriel Chilacatla." See also Eduardo Armendaris, *Estudios sobre las Aquas de Tehuacán*, pp. 8–9.

[8]León Marquez, "Breve Estudio," p. 36.

[9]Nahuatl soil, plant, and water terms are common even among non-Nahuatl speaking groups and suggest the antiquity of many traditional practices.

[10]Under Mexican law all natural resources, including water, are considered public goods and can be used or exploited only with the permission of the appropriate government agency. The formerly independent Secretaría de Recursos Hidráulicos (Hydraulic Resources) is now combined with the Secretaría de Agricultura y Ganadaría (Agriculture and Livestock).

Figure 9-1. Division of Texcali spring water (facing upstream), San Gabriel Chilac, Puebla, Mexico.

is still strong. Actual administration of Texcali water rests with the more than 900-member Junta de Aguas Federales de la Manantial La Taza, which allocates water to its members and represents them in negotiations with the government. Each member *(socio)* has one or more shares *(acciones)* of 6-hr flows in one or more of the four canals. Each *acción,* therefore, carries rights to water approximately every two months (900 ÷ 4 shares each 24 hr on each of 4 canals). The average allocation per share amounts to about (488 lps ÷ 2 villages ÷ 4 canals = ±.60 lps × 60 sec × 60 min × 6 hr =) 1.3 × 10^6 liters. Each user *(usario)* pays a small fee for his or her 6-hr period. The funds so generated pay for the expenses of the Junta.

Although one *acción* may be adequate to irrigate the many smaller, often less than hectare-size plots of Chilac, it is not enough for larger fields. Furthermore, the frequency of water allocation is unsuitable in this area of high evapotranspiration. Many farmers mention the need to irrigate every 15 to 20 days. To adjust need to availability an elaborate water market has developed for which I know of no equal in either the Old or New World.[11]

[11]Perhaps the most famous traditional water-exchange system has operated in Spain for hundreds of years. Thomas F. Glick, *Irrigation and Society in Medieval Valencia.*

The water market *(plaza de aguas)* convenes from 9:00 to 11:00 Monday through Friday mornings and from 2:00 to 4:00 Sunday afternoons[12] at a principal intersection in Chilac[13] near a spreading *cieba* tree.[14] At first only a few farmers appear; they form loose groups of three or four to discuss crops, weather, latest government regulations, prices in Mexico City, and other matters of interest. The conversations are in Spanish or Mexicana (Nahuatl); Chilac farmers have equal facility in either language.[15] The market is predominantly, but not exclusively, masculine; perhaps 10 percent of the crowd is women actively entering into discussions and negotiating for water. By 9:30 perhaps 100 farmers have arrived, forming large clusters around particularly interesting pieces of news, then dispersing into smaller groups for discussion and interpretation. Finally, about 9:45 water negotiations get under way.

Sellers move through the crowd announcing shares by specific Texcali canal or *galería,* time, terms, and usually a price but occasionally in trade for a different time. Buyers make counteroffers. If no agreement is reached, sellers move on, restating their position until they attract new potential buyers. When an agreement is reached, buyers receive scraps of paper (*cédulas,* certificates), on which are written details of the transactions. Similarly, buyers may circulate through the crowd announcing their water needs and soliciting responses from those with shares to sell or trade.

Prices for water vary enormously depending on precipitation and crop stages. Farmers are reluctant to discuss water quantities and prices in detail for commercial reasons and also because the practice is of questionable legality. In 1974 I heard peso equivalents of from US$8 to $56 for a six-hr share on one of the Texcali canals. These are substantial sums, especially for farmers without water, who might have to buy seven or eight *acciones* during a single cropping season.

[12]Apparently the market is convening at progressively later times. An old notice posted on a nearby *cieba* tree announces that the market convenes from 10:00 to 1:00 on Sundays and from 5:00 to 8:00 weekday mornings. A somewhat more recent sign on a building at the intersection advises that the market will be held from 7:00 to 9:00 weekday mornings and from 1:00 to 3:00 Sunday afternoons.

[13]While in session the market blocks the streets fairly well in fine disregard of a notice to the contrary posted on a building at the intersection.

[14]There is considerable symbolism in the *cieba* tree, which *galería* and other well diggers use as an indicator of underground water.

[15]The indigenous Nahuatl is almost universally called Mexicana in rural Mexico. Many Chilac farmers can also converse with Popolaca speakers from nearby hill villages who attend local periodic (not water) markets and work as laborers in the region.

Rivers

Although rivers offer more abundant and often more reliable flows than do runoffs or springs, the volume of water rivers carry makes management difficult. Only the most advanced traditional irrigation societies have possessed the power or the technology to dam rivers or store their waters.[16] A more common approach than dams is run-of-river systems, in which the river channels themselves are the storage component. The technical problem here is accessibility to water, especially during dry periods, when irrigation is most needed but stream levels are low.

The most common solution in Middle America is a low check dam *(presa)* that raises river levels only slightly but sufficiently to allow diversion into canals. An old and still widespread form is the *estaca* dam constructed by driving stakes *(estacas, palas)* into the river bottom in a line, either straight across the channel or angled upstream (Figure 9-2). Branches *(ramas)* or reeds *(carrizos)* are woven into the stakes, and the whole may be reinforced with rocks *(piedras)* and plastered with mud *(lodo)*.[17] More permanent cemented-rubble *(mampostería)* or concrete structures have replaced older *estaca* dams in some locations.

Irrigation water flows into canal networks through openings in the bank. Occasionally banks are simply breached to allow water to flow out, then repaired after the irrigation season, but this procedure creates dangerously weak spots. In most cases irrigation water flows out into canal networks through *mampostería* or concrete gates (*tomas; puertos*) in the banks. *Puertos* usually are blocked with boards or panels of sheet metal and sealed with mud. In improved systems steel gates are raised and lowered by long screws rotated by cranks or wheels.

In the simplest systems river water is diverted directly to streamside fields belonging to one or a few farmers, who take water from the river more or less at will. But often fields lie well downstream from diversion dams, and canal systems may serve many farmers in more than one village, giving rise to elaborate distribution and allocation systems. As an example, the three villages of Huitzo, Suchilquitongo, and Telixtlauca in the northern Valley of Oaxaca, Mexico, share a river-diversion system. Each farmer in the irrigated area has a right *(derecho)* to a

[16]In precolumbian times rivers descending to the Peruvian coast were dammed at upstream gorges to form lakes. The earth, rock, and masonry dams were built high enough "to hold all the descending waters." Ricardo E. Latcham, *La Agricultura Precolombiana en Chile y los Paises Vecinos.*

[17]The technique undoubtedly is precolumbian. Rabiela T. Rojas, *Aspectos Tecnológicos de las Obras Hidráulicas Coloniales en el Valle de México.* (Quoted in Carlos Olguin Palacios and Francisco Oyarzabal Tamargo, "Consideraciones sobre Riego y Drenaje Agrícola en la Prehistoria de México.")

Figure 9-2. *Estaca* diversion dam across Zahuapán River, Tlaxcala, Mexico. *Mampostería* take-off (*toma*) or gate (*puerto*) on far side.

standard 12-hr or fractional (for example, *medio,* half, or 6-hr) turn at a given frequency (for example, every 20 days). He may use his *derecho* or trade it as he wishes. A water judge *(juez de agua)* elected each year supervises allocations and mediates disputes.[18]

Water Sources Below Field Levels

Accessing water from lakes, swamps, and underground water tables poses special problems for traditional irrigators. The reason is clear: since lakes and swamps usually occupy the lowest point in an area, their waters are not available without lifting. Similarly, ground water must be raised at least to the level of plant root zones. Yet these water sources have been exploited intensively for centuries: lakes and swamps lie like drops of dew on the web of Middle American history and have played an integral part in the developing culture. Major civilizations rose and fell at lakeside locations, and to this day the region's orientation seems more lacustrine than maritime. Areas of high ground water also figure

[18]Susan H. Lees, *Sociopolitical Aspects of Canal Irrigation in the Valley of Oaxaca.* The *juez de agua* may be reelected repeatedly if he has managed to maintain order in the system without alienating too many participants.

prominently in historical and contemporary farming systems in Middle America.

How can farmers with severely limited energy resources lift the huge quantities of water needed for agriculture? Two solutions are directly associated with Middle America: manual irrigation, which concentrates on minimizing water use and is discussed here, and subirrigation, which is reviewed in Chapter 10.

Manual Irrigation

Manually lifting, transferring, and applying the quantities of water required for crop production are some of the most formidable tasks of nonindustrial farming. Nothing quite so epitomizes traditional drudgery as do the familiar images of farmers endlessly treading water ladders or of plodding oxen yoked to Persian wheels. If anything, the solutions in Middle America seem even less inviting. Without the wheel or draft animals, pre-Conquest farmers did not develop anything like the elaborate water-lifting devices found in the Old World,[19] and they have not adopted them widely since. Even counterpoise lifts, such as the shadoof, are curiously missing.[20] Lacking such devices farmers must transfer substantial masses of water without supplementary sources of energy or even mechanical devices. Yet manual irrigation has a long history in Middle America and remains important in several major agricultural districts. In the face of apparent disadvantages, we must look carefully at the systems themselves, and at their physical and cultural contexts, to discover reasons for their persistence.

Container Irrigation in the Central Valleys of Oaxaca

A series of reports from extensive archaeological research has made Oaxacan *riego a cántaro* perhaps the best known of all Middle American manual-irrigation systems.[21] In those areas on the valley floor

[19]Aldert Molenaar, *Water Lifting Devices for Irrigation.*

[20]A few implements introduced by the Spanish, such as the wheel or chain of buckets *(noria)* and windmill, have enjoyed limited acceptance. But one can only share George M. Foster's (*Culture and Conquest,* pp. 62–68) puzzlement over the fact that the abundant waters and adjacent arid lands of America did not promote more general use of these ingenious devices.

[21]Kent V. Flannery et al., "Farming Systems and Political Growth in Ancient Oaxaca"; Anne V. T. Kirkby, *The Use of Land and Water Resources in the Past and Present Valley of Oaxaca, Mexico.* The phrase *riego a/con/de cántaro* is commonly used by Oaxacan farmers and serves better than the more general "pot irrigation" proposed by Flannery et al. and Kirkby to distinguish

where water tables are less than 10 m below the surface farmers open small wells in the fields and draw water up by hand using special clay or metal vessels *(cántaros),* then carry them along narrow paths and carefully pour them into round planting pits or rectangular ridged beds. The system is laborious and time consuming. It also makes remarkably efficient use of water and is the basis for a successful agricultural industry. Most of the vegetables, herbs, and flowers sold in the huge regional market in Oaxaca City and in the smaller markets at Ocotlán, Zaachila, and Zimatlán come from *cántaro*-irrigated fields.

The Oaxaca Valley is dry (Table 9-1), with barely enough precipitation for a single crop during the short rainy season, and even then moisture-conserving measures are necessary.[22] Much of the valley is underlain by impermeable layers of base rock that restrict subsurface storage to relatively thin layers of alluvium. High or perched water tables occur in the central section, where the three valleys that form the Oaxaca Valley join. A variety of water-management systems have been employed here since ancient times, including flood-water, canal, and well irrigation, and subirrigation. Where water tables are within a meter or so of the surface, a form of subirrigation, or drained-field (also, water-table), farming is possible. Where the phreatic (water-table) surface is lower, water is extracted by hand from shallow wells with containers and carried to planting beds.

Fields and Crops

Oaxacan *riego a cántaro* fields range in size from plots of a few square meters up to about 0.25 ha. Initial plowing with oxen is followed by careful hoe and hand preparation of field surfaces. Animal manures, chemical fertilizers, leaf and crop residues, and in some parts of the valley debris from ants' nests enrich the soil and improve tilth and moisture absorption and retention.

Seedlings of large plants, such as tomatoes and chile peppers, are transplanted from seed beds into circular ridged pits *(cepas)* up to 90 cm in diameter, aligned in neat rows (Figure 9-3). Planting pits are

---

Oaxacan practice from other container systems discussed here. José L. Lorenzo's ("Aspectors Físicos del Valle de Oaxaca") *riego a brazo* (arm irrigation) is a descriptive phrase that could be applied to any type of manual irrigation. Pot irrigation more accurately refers to the ancient Southwest Asian practice of burying water pots next to plants, a technique that approximates modern trickle irrigation. M. H. Alemi, "Distribution of Water and Salt in Soil Under Trickle and Pot Irrigation Regimes"; R. C. Mondal, "Farming with a Pitcher."

[22]Bruno V. Ferrari Bono, "Desarrollo de los Recursos Hidráulicos en los Valles Centrales de Oaxaca"; Lorenzo, "Aspectos Físicos"; Enriqueta García et al., *Precipitación y Probabilidad de Lluvia en el Estado de Oaxaca y su Evaluación;* Kirkby, *The Use of Land and Water Resources,* pp. 35-36.

Table 9-1. Temperature, Precipitation, and Potential Evapotranspiration for Three Cities in Mexico

	Jan.	Feb.	Mar.	Apr.	May	Jun.	Jul.	Aug.	Sep.	Oct.	Nov.	Dec.	Totals
Oaxaca City (17° 04′N, 1,563 m elevation)													
Temperature (T), °C	18.4	19.7	21.9	23.3	23.2	21.9	21.1	21.3	20.7	19.8	19.1	18.3	$\overline{X}$ = 20.7
Precipitation (P), mm	3	3	6	27	68	134	84[a]	96[a]	123	48	6	4	602
Potential evapotranspiration (PET), mm	58	63	93	107	114	98	92	92	80	69	59	58	983
P - PET	-55	-60	-87	-80	-46	36	-8	4	43	-21	-53	-54	-381
Puebla City (*Tlaxcala*) (19° 02′N, 2,150 m elevation)													
Temperature (T), °C	14.0	15.7	17.9	19.2	19.1	18.5	17.7	18.0	17.3	16.7	15.7	14.3	$\overline{X}$ = 17.0
Precipitation (P), mm	8	6	5	28	71	161	136[a]	155	164	72	23	6	835
Potential evapotranspiration (PET), mm	43	49	71	82	86	79	79	76	64	60	50	43	782
P - PET	-35	-43	-66	-54	-15	82	57	79	100	12	-27	-37	53
Mexico City (*Chinampa* district) (19° 24′N, 2,308 m elevation)													
Temperature (T), °C	12.5	14.4	16.7	17.6	17.6	17.3	16.1	16.3	15.7	15.0	13.7	12.6	$\overline{X}$ = 15.5
Precipitation (P), mm	9	3	8	24	55	111	152	133	120	49	21	7	692
Potential evapotranspiration (PET), mm	40	49	65	72	78	76	68	66	61	57	45	40	717
P - PET	-31	-46	-57	-48	-23	35	84	67	59	-8	-24	-33	-25

SOURCES: Basic climate data from the Files of Servicio Meterologico Nacional, Tacubaya, D.F., Mexico. Potential evapotranspiration calculated according to C. W. Thornthwaite and J. R. Mather, *Instructions and Tables for Computing Potential Evapotranspiration and the Water Balance.*

[a]Precipitation decreases in July and August are the result of recurrent midsummer droughts.

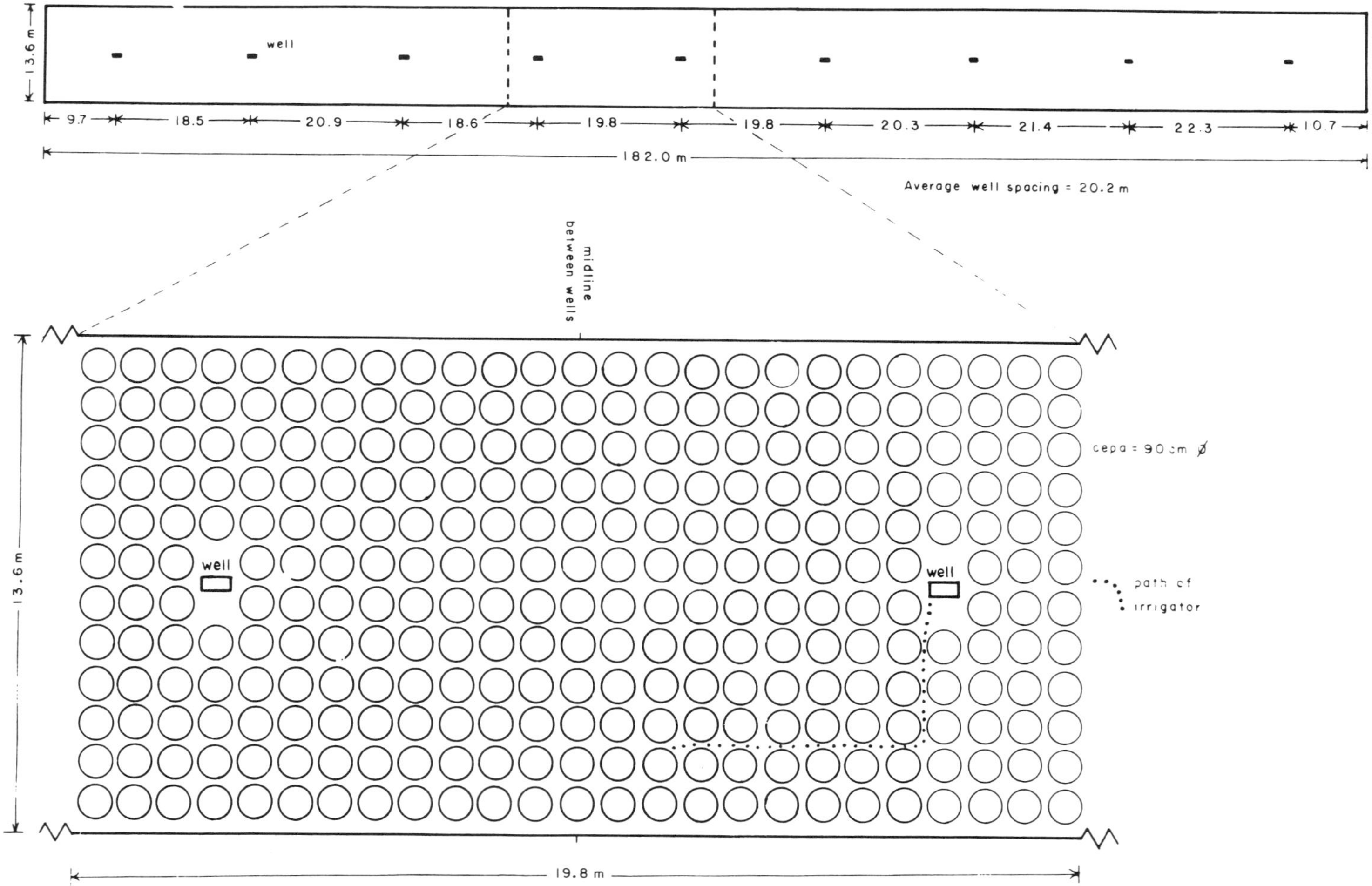

Figure 9-3. Field of *cepas* manually irrigated by *cántaro*, Zaachila, Oaxaca, Mexico.

separated by 10- to 30-cm spaces, which allow access to the plants. Thus, only a fraction of the total field surface is irrigated.

Smaller plants such as onions, garlic, or herbs are transplanted or seeded directly into square or oblong, ridged planting beds *(cajetes)*[23] 50 cm to a side but often less. *Cajetes* are precisely formed using a string *(hilo)* for perfect alignment and close hoe and hand work to form tiny borders. Several lines of *cajetes* make up a planting surface known by variety of names, such as *canoa* (literally, canoe), *melga* (from *amelga,* a plot of land between furrows or ridges), and *tablón* (Figure 9-4 and Figure 7-16, page 145). Because little space is lost to paths and borders, more than half of most *cajete* plots is irrigated. Plants are carefully spaced within *cajetes,* usually in multiples or fractions of a handspan *(cuarta),* or the distance between extended tips of the thumb and little finger, which averages about 20 cm. For example, onions are planted two per *cuarta,* or at about 10-cm intervals.

Because *cepas* and *cajetes* are precisely and geometrically arranged in the long planting beds, the total number of plants in a plot can be quickly calculated. In addition, the tiny plots are subject to almost daily inspection during meticulous hand cultivation and watering. Individual plant losses are noted, and it is not uncommon for farmers to know precisely how many plants their plots contain.

Water Sources

Farmers excavate shallow, midfield wells with shovels and spades, and lift out the dirt in baskets. Wells are narrow, from 40 to 50 cm wide and 80 to 100 cm long, the bare minimum within which a person can work. Well depth is determined by distance to ground water, usually extending a meter or so below the water table to allow for seasonal fluctuations and draw-down during extraction. I measured wells from 2.5 to 9 m deep, and heard of others 12 to 13 m deep. But 10 m appears to be about the acceptable limit because of the difficulty of drawing water from deeper wells. In loose soils a log or two laid across the well opening serves as a foot rest and reinforces the edge. Baskets imbedded in well bottoms reduce caving and silting.

Kirkby reports that wells are filled in and fields changed every 1 to 3 yr despite the labor involved in digging new wells; she suggests that the threat of salinization may account for this rapid rotation.[24] Fields are not fallowed but revert to *temporal* (rain-fed) maize cultivation. From my observations, however, the length of time plots are manually irrigated is quite variable and depends on such factors as the amount of

[23]*Cajete,* from the Nahuatl *caxitl,* a flat earthen bowl; by extension, a planting bed hollowed out or rimmed with small borders.

[24]Kirkby, *The Use of Land and Water Resources,* pp. 42, 100.

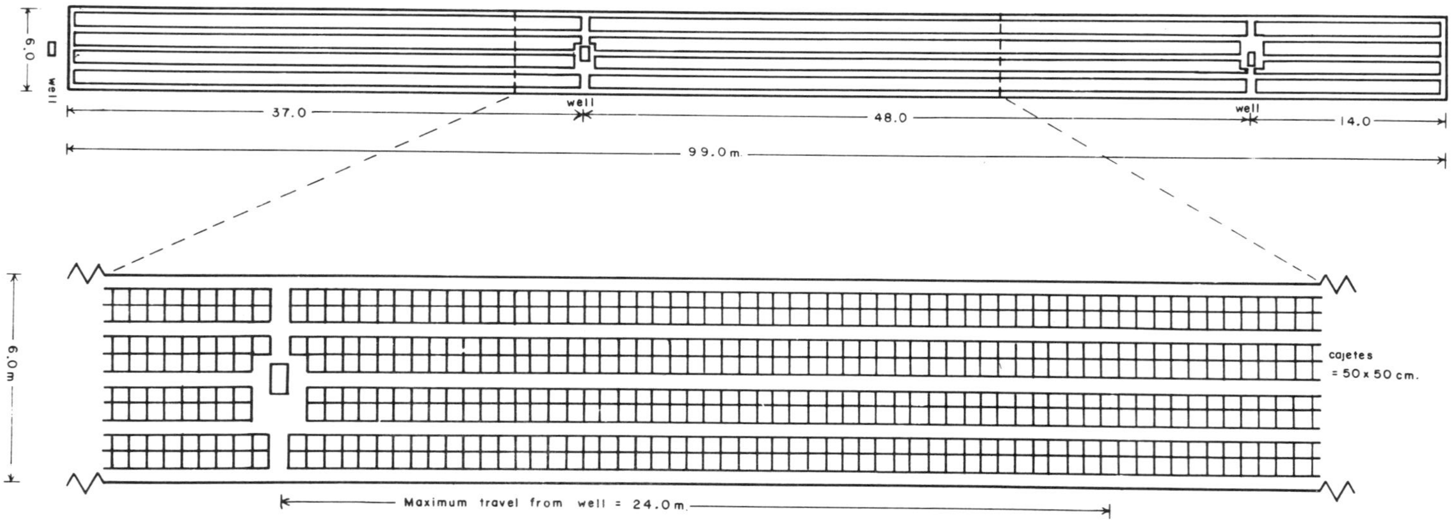

Figure 9-4. Field of *cajetes*, San Antonino, Oaxaca, Mexico.

irrigable land a farmer holds, the depth of (and thus the amount labor invested in) wells, and the likelihood of salt accumulation. Some fields are hand irrigated for many years.

Space between wells is based on the need to minimize several factors: the distance water has to be carried, excavation effort, and field space taken up by wells. Wells are spaced at intervals of 20 to 30 m and never more than 40 m along the central line of a field. Because plots are narrow, from 5 to 15 m wide, maximum distances from wells to crops are about 15 to 20 m and average distances are half of that (Figures 9-3 and 9-4).

Cántaros

Cántaros[25] used in the Oaxaca Valley vary in capacity from 10 to 14 liters and are of a design found in many parts of the world. From a strongly flared lip and constricted throat 10 to 13 cm in diameter, the vessel swells to a three-quarter ovaloid body with an equatorial diameter of 28 to 32 cm, then narrows to a rounded point at the base. The distinctive form makes the *cántaros* self-filling even in narrow wells because they pivot around the wide equator and will not float even when partially filled. *Cántaros* are made of black earthenware, produced by the reduced firing technique in such famous pottery centers as San Bartola Coyotepec, or are fabricated of metal in small soldering shops in Oaxaca City or Ocotlán (Figure 9-5).

Irrigation Method

In a typical operation farmers drop the self-filling container down a well, letting the rope, usually of *maguey* fiber *(ixtle),* run through their hands. The rope, which is tied below the flared lip, is long enough to reach the bottom of the well with an additional meter or so that is sometimes tied around the farmer's waist. As soon as the vessel fills, the farmer retrieves it hand over hand, coiling the rope, which is held in the left hand or looped over the shoulder (Figure 9-6). Carrying the *cántaro* by the flared lip, the farmer hurries down paths between planting pits or beds. Access paths are at right angles to each other because the geometric pit or bed arrangements and the delicate formation of plot surfaces prohibit walking along diagonals or random routes. Thus, to water a plant 10 m down field and 5 m to the side of a well requires a round trip of 30 m rather than the 22.36 m of direct distance.

[25]The Latin *cantharus* (Greek *kantharos*), from which the term *cántaro* is derived, was a deep, two-handled drinking cup. The Oaxacan vessels more closely resemble a handleless *hydria* or *kalpis,* which were used for drawing water. Edwin Atlee Barber, *the Ceramic Collector's Glossary;* Robert S. Folsom, *Handbook of Greek Pottery.*

Figure 9-5. Earthenware and metal *cántaros*, Oaxaca, Mexico.

Arriving at a planting bed or pit, the farmer shifts the *cántaro* so that the right hand supports the neck while the left hand tips up the pointed bottom. Free fingers of the right hand break the flow, and the water falls gently on the plants and soil surface (Figure 9-7). In seed beds and when plants are young, the right hand is agitated to produce an even finer spray in order to avoid plant damage and splash erosion. In the few seconds it takes to discharge a *cántaro* the farmer examines each plant in the pit or bed for parasites, physical condition, development, and so on.

Farmers keep careful watch on plant and soil conditions and irrigate as needed. During the irrigation season, water is applied every 2 or 3 days. Infrequent showers provide welcome respite from the drudgery of hand watering. If evaporation rates are high, however, it may be necessary to irrigate daily. Thus, the response to day-to-day variation in water needs is frequency of watering, not quantities applied, and farmers must be prepared to irrigate their plots completely in one day.

Amounts of water applied vary with stages of plant growth and with season. For example, when tomatoes are young and evaporation rates are relatively low, three pits are watered with one *cántaro* (12 liters). Later in the season one *cántaro* supplies only two pits. Thus, in

Figure 9-6. Lifting filled *cántaro*. Well opening is directly below the container.

the field illustrated in Figure 9-3 a circle about 50 cm in diameter is wetted in each *cepa*. Allowing for a few additional centimeters of horizontal movement, I calculate that the contents of each 12-liter (12,000 cc) *cántaro* cover about 0.8 m^2 in the three *cepas* for a depth of 1.5 cm. Later in the season, when only two *cepas* are irrigated with one *cántaro,* the application depth is 2.25 cm. Similarly, early in the season one *cántaro* will satisfy three of the 50-×-50-cm *cajetes* illustrated in Figure 9-4, but only two later on. Assuming that the small borders are also wetted by splashing or horizontal movement, the smaller 11-liter (11,000 cc) *cántaro* used for *cajetes* provides about the same depths of water. It is interesting to compare water application with evapotranspiration rates (Table 9-1). As Kirkby notes, Oaxacan farmers apply just enough water to offset losses and provide a little extra for drainage.[26]

[26]Kirkby, *The Use of Land and Water Resources,* p. 42.

Figure 9-7. Discharging *cántaro* into *cepa* of tomato plants, Zaachila, Oaxaca, Mexico.

Labor Requirements

Any manual procedure used by many individuals with different soils and crops is bound to comprise substantial variations in production rates. However, my own and Kirkby's data are comparable and also are consistent with calculated evapotranspiration losses and irrigation needs. The task times reported here are averages from a series of measurements I took in the Zaachila Valley of Oaxaca. The figures reveal the amount of work this irrigation method requires. They also reveal the close relationship of biological and cultural elements; crops, weather, labor, and field size are all intimately linked through the mechanics of manual irrigation.

Cántaros are dropped and lifted at the rate of about 1 sec/m. An 11- or 12-liter container fills in about 9 seconds, providing the well contains sufficient water.[27] Carrying rates from wells to plants and back average 2 sec/m. Discharging a *cántaro* takes about 12 seconds.

In a field where average well depth is 7 meters and average distance from wells to plants is 7.5 m, the water delivery rate would be (7 + 9 + 15 + 12 =) 43 seconds per 12-liter *cántaro,* or about 17 liters per minute.[28] In six hours of continuous work at this rate, farmers can apply 15 mm of water to 400 m² of plot surface.

But *riego a cántaro* is arduous; no man can work continuously throughout a day without stopping. My observations suggest that about 25 percent of an irrigator's time is spent in short rest and meal periods. Thus, applying 15 mm of water to 400 m² actually takes about 8 hours.

As the season progresses evapotranspiration rates and plant water needs increase. The popular sizes of *cántaro* are the largest that a man can work with on a sustained basis. Extraction and travel times cannot be reduced significantly. That leaves length of working day as the only variable that can be adjusted. As the depth of water applied increases from 15 to 20 or 25 mm, the 400 m² plot requires 8 to 10 hours of continuous irrigation, and working days lengthen to 10 or 13 hours, or about the maximum possible for work of this nature.[29]

Given climate, plant, soil, and water requirements, and the limits to human energy, it would be difficult for an individual farmer to increase the amount of land farmed under this form of irrigation.[30] The familiar relationship between farming system and farm size is nowhere better illustrated than when the principal power source is the farmer himself!

Calculating total water transfers for a season will give some feeling for the work involved in container irrigation. To water the field shown in Figure 9-3 every third day, allowing one *cántaro* for three *cepas* early in the four-month season and two later on (Table 9-2), the farmer would cover well over 500 km, half of that carrying a full *cántaro* and rope that together weigh about 15 kg. On the same basis, irrigating the *cajete* field shown in Figure 9-4 would require walking about 600 km and lifting

[27]Not infrequently wells cannot produce enough to keep up with even hand irrigation, and some can be worked for only a few hours at a time. Kirkby, *The Use of Land and Water Resources,* p. 24. Under these circumstances *cántaros* fill only partially and extraction rates fall off sharply.

[28]The low delivery rates in manual irrigation are best expressed in liters per minute (lpm) rather than the more common liters per second (lps), even though individual steps in the irrigation cycle were measured in seconds.

[29]Temperatures of 27°C in the shade were recorded during these observations, and there is seldom shade in *cántaro*-irrigated fields.

[30]The plots shown in Figures 9-3 and 9-4 are both somewhat larger than would be expected from the calculations. The *cepa* field was worked by a young Zapotec farmer proud of his strength and quickness who boasted that he could irrigate more land than most. The *cajete* farmer could count on help from a teenage son.

Table 9-2. Schedule for Tomatoes, Zaachila, Oaxaca

Farming operation	Irrigation	Date
Seed bed planted		1 Nov.
	Careful splash from *cántaro*	15 Nov.
Transplanting		15 Dec.
	Cántaro irrigation	15 Jan.
		15 Feb.
		15 Mar.
Harvest begins		15 Apr.
		15 May
	Temporal season begins	

more than 300 m tons of water from relatively deep 6.5 m wells. Expressed in these terms the labor requirements of *cántaro* irrigation assume heroic proportions.[31]

Gasoline or electric pumps seem an obvious answer to the unending toil of *cántaro* irrigation. There are two obstacles to the widespread use of pumps, however, one hydrologic and the other socioeconomic. It seems doubtful that the shallow Oaxacan aquifers could sustain much pump extraction. Those affluent farmers already owning small 1- to 2-in. pumps often find that they can operate them for only a few hours at a time because of insufficient subterranean flow. Pumps may lower the water level in an area so much that *cántaro* irrigation is impossible. In addition, because pump-extracted water is delivered to plots in furrows, it is less efficiently applied than the water carefully carried and poured in *cántaros*. As Lees points out, for 3,000 years hand irrigation "indirectly regulated the use of water and maintained a long-term balance between farmers and their resources."[32] Increasing use of pumps will disrupt this balance to the detriment especially of the small irrigators.

If limited ground-water supplies restrict pump extraction, it would be difficult for an individual farmer to expand his irrigated area sufficiently to make optimum use of a pump. Yet even if plot sizes were not increased the investment would seem justified on labor savings

[31]Similar rates are achieved in other regions of the world. Some human-powered paddle wheels in Asia, for example, lift but 208 lpm. Molenaar, *Water Lifting Devices*. To flood a 2,500 m² field with 15 mm of water would require 18 hr of steady treading, the equivalent of quite a few kilometers.

[32]Lees, "Oaxaca's Spiraling Race for Water," p. 30.

alone. In reply to this suggestion many farmers complain that they simply do not have enough money to make the initial investment in pumps and hoses. Joint use of equipment would seem a solution and is being tried in several villages. But cooperative pump operation has its own problems, especially in scheduling and maintenance, and would be especially difficult in Oaxaca, where the highly individualistic *riego a cántaro* has not established a tradition of cooperative irrigation.

Whether the laborious *riego a cántaro* appears at all feasible to Western eyes, the fact remains that it has endured for 3,000 years and is still a viable and productive system. Although costs in human energy are high and limitations in human energy restrict field size and farmer income, it is one solution to the problem of irrigating where surface water is not available and ground water is scarce but relatively close at hand. Despite its many disadvantages, one United Nations expert has suggested expanding the system in Oaxaca.[33]

Other Container Regions

No other container-irrigation system in Middle America can rival the Oaxacan *riego a cántaro* in its longevity, intensity of use, and specialized technology. It is, however, but one example of a family of irrigation practices still found in many places where physical and economic conditions favor its use. One such region is southwestern Tlaxcala, Mexico, where *riego a bote* (literally, to irrigate by pot or jar) was once quite common but is found now only occasionally.[34] (Precipitation, temperature, and evapotranspiration data for this region are given in Table 9-1.)

Tlaxcalan *riego a bote* fields are as long and narrow as the *cántaro* fields in Oaxaca, but individual planting beds tend to be slightly larger. Many fields are bordered by straight-sided drainage ditches, which are convenient sources of irrigation water, as are shallow wells only 1 to 3 m deep, which can be dug almost anywhere on the basin floor. Wells are larger than those in Oaxaca, on the order of 1 to 2 m wide and long, to accommodate the larger containers, which are not self-filling.

The most popular container in Tlaxcala is the familiar can from which cooking grease *(manteca)* is sold in country stores. When new, the cans measure 24 cm on each side and are 35 cm deep. A piece of wood nailed across the opening serves as a handle. Gross capacity when new

[33] Ferrari Bono, "Desarrollo de los Recursos Hidráulicos." p. 15.

[34] Because *riego a bote* is no longer popular in Tlaxcala, examples are scarce and contemporary data on well sizes, extraction rates, and field dimensions may not be accurate for the former, more widespread practice. Information presented here is from field observations and from accounts of farmers regarding manual irrigation in times past.

is 20 liters, but deformation from use and spillage in transit reduce effective carrying capacity to about 18 liters. Other containers, such as buckets and pots of various sizes and designs, are also pressed into service.

In general, Tlaxcalan *riego a bote* is similar to Oaxacan *cántaro* irrigation. The *bote,* secured by a rope tied to the wooden handle, is cast into a *zanja* (ditch) or well with a slight dragging motion to hasten filling. The farmer then retrieves the container and hurries (they use the verb *correr,* to run) along narrow paths to planting beds. Water is swirled to break the force over an area of about a square meter (Figure 9-8). Alternatively, water may be poured into furrows. Although this procedure results in a less efficient use of water, it is considerably faster because the irrigator transfers water only from source to furrow rather than to all parts of the plot.[35]

A third region of extensive container irrigation is the famous *chinampa* district of the Basin of Mexico. (Temperature, precipitation, and evapotranspiration data are given in Table 9-1.) When *chinampas* are constructed, the process simultaneously creates both fields and canals. The canals range in size from main aquatic thoroughfares 50 m across to channels no more than a meter wide alongside individual plots. Access to the plots is by boat *(canoa),* another of the distinctive features of the system.

Because water readily infiltrates the porous *chinampa* soils, water below the plots is at the same level as in the adjacent canals. According to popular account, capillary transfers from water table to root zone provide adequate moisture for cultivation. But although some crops benefit from this supply, it falls far short of being a complete irrigation system. In true subirrigation (as is explained in Chapter 10), farmers usually have to adjust water levels so that water is continuously in root zones,[36] and *chinampa* farmers *(chinamperos)* lack this degree of control over water supplies from canals.

An alternative method of irrigation is needed for shallow-rooted crops and for times when canal levels are low. Although a variety of manually operated water-lifting and water-application devices are mentioned in the literature as having been used in the past, including scoops, shovels, oars, buckets, pots, and combinations such as pole and

[35]Some Oaxacan farmers also mentioned pouring *cántaros* directly into furrows.

[36]Food and Agriculture Organization and United Nations Educational, Scientific, and Cultural Organization, *International Source-Book on Irrigation and Drainage of Arid Lands in Relation to Salinity and Alkalinity,* pp. 430–31. For other misconceptions about *chinampa* construction and operation, see Gene C. Wilken, "A Note on Buoyancy and Other Dubious Characteristics of the 'Floating' Chinampas of Mexico."

Figure 9-8. Discharging *bote* over onion seedings, Tlaxcala, Mexico.

bucket,[37] contemporary *chinampa* morphology generally favors containers over other devices.

The use of scoops or shovels seems dubious. *Chinampa* plot banks drop off steeply, leaving no shoal area on which to stand. If irrigators used scoops or shovels, they would have to stand on the plots themselves and reach down to the canal, at best a difficult working position and difficult motion to make. Alternatively, *chinamperos* could stand in their *canoas* and splash water onto plot surfaces with scoops, shovels, or even oars. On plots no more than a few decimeters above canal levels this procedure might be feasible. But for higher fields, it would mean lifting water more than a meter and applying it over a surface several meters wide. Considering the shallow draft and narrow beam of the *canoas,* and the force necessary to lift and impel significant quantities of water, one wonders how *chinamperos* would keep themselves balanced.

Some writers suggest that the same long-handled scoop *(zoquimaitl)* used for dredging muck from canal bottoms was also used for irrigation. Thus, Price says that water might have been "supplied as necessary by

[37]Pedro Armillas, "Gardens on Swamps"; Pedro Armillas, "Notas sobre Sistemas de Cultivos en Mesoamérica"; Michael D. Coe, "The Chinampas of Mexico"; Elisabeth Schilling, *Die "Schwimmenden Gärten" von Xochimilco;* Robert C. West and Pedro Armillas, "Las Chinampas de México."

the cultivator from his boat with a long-handled scoop."[38] Because irrigation water is no heavier than the saturated muck used for field dressing, the procedure of lifting water from canals a meter or so below plot levels and then spreading or splashing it over plot surfaces 6 to 8 m wide with a scoop attached to the end of a 2- to 3-m pole is technically possible. However, it seems an awkward and inefficient method of applying irrigation water.

The case for container irrigation is strong and is reinforced by contemporary *chinampero* practice. From plot edges, buckets, lard cans, and other containers are cast into canals, then discharged with a sweeping spray pattern over plot surfaces. Except that farmers must lean out over canals to retrieve containers, water extraction is no more arduous than lifting from wells. Generally *chinampa* container irrigation goes faster than Oaxacan *riego a cántaro,* primarily because the water is close at hand at no great depth and the containers are larger. Some *chinamperos* have mounted small pumps in their boats (Figure 9-9). Water pumped from canals is forced through nozzles to produce a spray much like that which can be applied by hand. In contrast, Oaxacan and Tlaxcalan farmers who adopt pumps usually switch to furrow irrigation.

Scattered reports from other regions suggest that container irrigation is relatively widespread in contemporary Middle America. In villages around Lake Atitlán, Guatemala, tomatoes grown in large pitted mounds are irrigated every third day with lake water carried in special jars *(tinajas).* Although ordinarily these are women's vessels, both men and women irrigate with them (Figure 9-10).[39] The practice is especially interesting because this region is best known for scoop, rather than container, irrigation.[40]

In El Salvador farmers use a variety of metal and plastic containers and sprinkling cans to irrigate tiny plots of vegetables. Water is scooped directly out of small streams and carried to flood-plain fields, where it is carefully splashed over such crops as tomatoes, radishes, and carrots. The vegetables are sold in nearby San Salvador. Part of the proceeds are used to pay rent for the tiny plots.[41]

Scoop Irrigation with Shovels

In the second group of manual-irrigation systems, water is scooped or splashed directly onto fields from small canals or drains with scoops,

[38]Barbara J. Price, "Pre-Hispanic Irrigation Agriculture in Nuclear America," p. 22.

[39]Felix Webster McBryde, *Cultural and Historical Geography of Southwest Guatemala,* p. 26.

[40]Kent Mathewson, *Irrigation Horticulture in Highland Guatemala;* pp. 93-96.

[41]Albert G. Madsen, personal communication.

Figure 9-9. Pump mounted in canoe drawing water from canal and discharging it onto *chinampa* plot surface, San Gregorio Atlapulco, Mexico.

Figure 9-10. Lake-side fields, Santiago Atitlán, Guatemala.

basins, trays, or gourds. Although in some cases water may be lifted from sources as deep as a meter, most scoop systems use water standing or flowing just below field level. In fact, some of the scoop systems reviewed here could be easily converted to flood or furrow irrigation. But strong and probably well-founded belief in the efficacy of scoop-splashed irrigation, suitable environmental conditions, and wage rates that permit extensive use of human labor account for the persistence of these practices.

Because water cannot be scooped or thrown in any volume or with any accuracy more than about 10 m, it must be available immediately adjacent to planting surfaces. Thereafter the process is straightforward: farmers determine moisture requirements with a quick appraisal of soil conditions, plant needs, and atmospheric conditions, then toss or splash water directly over planting beds until light surface puddling occurs. The process is repeated as often as necessary: daily when evaporation is high; less frequently when water losses are low.

Like container irrigation, watering with scoops has the flavor of ancient practice, although pre-Spanish precedents have not been documented. Present scoop-irrigation regions are highly localized, limited in extent, and associated with particular crops or culture groups or both. Evidence from other areas, however, suggests wider distribution in Middle America in times past than now.

The close but distinct farming areas of Almolonga, La Ciénaga, and Zunil in the department of Quezaltenango, Guatemala, use similar scoop-irrigation systems (called *riego con pala*), which are discussed together here. Almolonga and Zunil occupy separate small valleys 5 and 9 km respectively southeast of Quezaltenango City, the capital city of the department. La Ciénaga is part of Quezaltenango, occupying a low area on the northeast edge of the urban area. Guatemalan scoop-irrigation areas are not as dry as the container areas of Mexico (Table 9-3). Nevertheless, a pronounced dry season from November to May makes irrigation mandatory for year-round cropping and advisable on a stand-by basis even during the rainy months.[42]

Fields and Crops

The basic land unit in Quezaltenango is the *cuerda,* measuring 25 *varas* (1 *vara* = 0.84 m) to a side, or 625 *varas*^{2} (about 440 m^2). Manually irrigated *tablones* at Almolonga and La Ciénaga are uniformly 12.5 *varas* (10.5 m) wide and vary in length from 20 to more than 100 m with canals or drains on each side (Figure 9-11). Thus each *tablón* is one to several *cuerdas.* An individual farm may consist of from a fraction of a

[42]José Luís Rosales Padilla, *Evaluación de Sequías en la República de Guatemala con Base en Datos de Precipitación Mensual.*

Table 9-3. Temperature, Precipitation, and Potential Evapotranspiration for Three Cities in Guatemala

	Jan.	Feb.	Mar.	Apr.	May	Jun.	Jul.	Aug.	Sep.	Oct.	Nov.	Dec.	Totals
Quezaltenango City (14° 50′N, 2,333 m elevation)													
Temperature (T), °C	12.0	13.1	14.4	16.6	17.2	16.6	14.4	17.3	16.2	15.3	14.5	12.8	$\overline{X}$ = 15.0
Precipitation (P), mm	1	2	0	11	30	136	234	167	230	91	10	3	915
Potential evapotranspiration (PET), mm	41	41	56	66	77	68	60	75	64	58	51	44	701
P - PET	-40	-39	-56	-55	-47	68	174	92	166	33	-41	-41	214
Solola City (14° 46′N, 2,114 m elevation)													
Temperature (T), °C	14.3	15.2	18.3	19.4	21.4	19.7	20.0	19.8	20.4	18.8	17.7	14.9	$\overline{X}$ = 18.3
Precipitation (P), mm	0	1	4	20	50	289	285	309	537	187	36	0	1,718
Potential evapotranspiration (PET), mm	38	44	68	78	100	81	87	81	86	70	60	44	837
P - PET	-38	-43	-64	-58	-50	208	198	228	451	117	-24	-44	881
Huehuetenango City (15° 19′N, 1,902 m elevation)													
Temperature (T), °C	15.2	16.0	17.4	18.5	18.5	17.6	17.5	17.6	17.2	16.7	15.7	15.3	$\overline{X}$ = 16.9
Precipitation (P), mm	0	2	12	20	116	185	157	118	189	226	11	0	1,036
Potential evapotranspiration (PET), mm	52	55	71	78	83	75	77	75	67	64	54	52	803
P - PET	-52	-53	-59	-58	33	110	80	43	122	162	-43	-52	233

SOURCES: Basic climate data from Observatorio Nacional (Guatemala), *Datos Meteorológicos de las Cabeceras Departamentales.* These 1972 data, however, are identical to those published in Observatorio Nacional (Guatemala), *Datos Meteorológicos Mensuales Hasta 1959 Inclusiva,* in 1968 and thus do not reflect current or particularly long-term observations. Potential evapotranspiration calculated according to C. W. Thornthwaite and J. R. Mather, *Instructions and Tables for Computing Potential Evapotranspiration and the Water Balance.*

Figure 9-11. Fields at La Ciénaga, Quezaltenango, Guatemala, showing *tablones* and bordering *zanjas*, from which irrigation water is scooped.

cuerda to 5 or 6 *cuerdas*, with the average perhaps 3 or 4 *cuerdas* per *dueño*.[43] Holdings may be contiguous on the same *tablón* or scattered over several. Plots are not divided into smaller units, such as the *cepas* or *cajetes* of Oaxaca, nor are they edged with *bordos*. *Tablones* at Zunil are smaller and less regular in shape than are those in Almolonga and La Ciénaga and are stepped down in a series of small terraces, some of which are faced with rock. Canals are smaller than at Almolonga or La Ciénaga and also are often rock lined (Figure 9-12).

Water Sources

The canals or *zanjas* bordering each *tablón* vary in size from the large, meter-wide drains of Almolonga and La Ciénaga to the tiny canals of Zunil, which are only a few centimeters wide and deep. At

[43]Although William C. Merrill, Lehman B. Fletcher, and Michael S. Hanrahan ("Vegetable Production and Marketing in Guatemala") show average farm size in Quezaltenango as 0.2 ha, or about 5 *cuerdas* (p. 19, Table 2.2), they later note (p. 30) that landholdings in Almolonga are usually only a fraction of a *cuerda* and that the largest known vegetable producer farms 7 *cuerdas* (0.3 ha). In addition to tiny produce plots on the valley floor, farmers work larger *temporal* fields in the adjacent hills, where they grow subsistence crops such as maize and beans.

Figure 9-12. Tiny plots and rock-lined canals, Zunil, Quezaltenango, Guatemala.

regular intervals along each waterway there are small pools, often no more than slightly widened places, from which water can be extracted.

Water is occasionally in short supply. Low water impedes scoop transfers and slows rates at which water can be delivered to fields. Thus, the water shovel used at Almolonga and La Ciénaga does not deliver a full scoop when water in the extraction pools drops below 10 cm, and the Zunil scoop loses efficiency below 5 cm. During these infrequent periods farmers adopt irrigation schedules to relieve pressure on the low flows. But I have heard no reports of shortages serious enough to require assigning priorities or rescheduling crops.

Scoop Types

At Almolonga and La Ciénaga water is extracted from drains and canals with an appropriately named *pala de agua* or *de riego* (water or irrigation shovel or scoop), which is an oblong, shallow, wooden scoop about 29 cm long, 17 cm wide, and 7 cm deep, with the front angled back at about 45°. A wooden handle perhaps 130 cm long and 3 cm in diameter is nailed to the bottom of the scoop (Figure 9-13). *Palas* are made entirely of local pine from 5 to 10 mm thick, or they may have thin galvanized metal bottoms (Figure 9-14). The Zunil *pala* (Figures 9-13 and 9-14) consists of a shallow galvanized metal cone about 25 cm

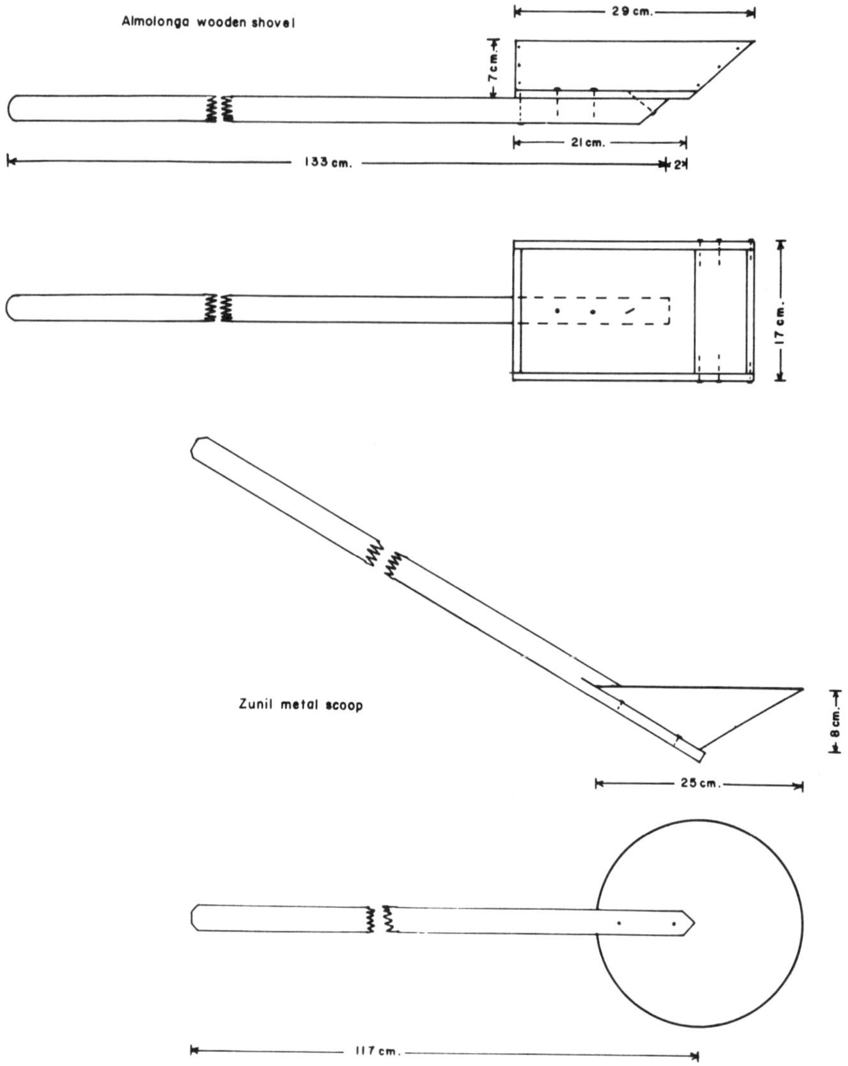

Figure 9-13. *Palas de riego,* Quezaltenango, Guatemala.

in diameter and 8 cm deep at the vertex, mounted on a wooden handle that is slightly shorter than the Almolonga handle. The *palas* are made by local carpenters in Almolonga and Zunil.

Irrigation Method

In use the Almolonga and La Ciénaga *pala de riego* is held with both hands and swung much like an ordinary shovel. Water is scooped

Figure 9-14. Wood Almolonga water shovels with metal bottoms (top), and Zunil conical scoop (bottom). Hat for scale.

up with the angled lip, then tossed with a continuous sweeping motion as the shovel is rotated slightly to improve accuracy and control spray patterns (Figure 9-15). Alternatively, the handle may be placed under the armpit and braced behind the shoulder so that the scoop can be levered with a series of rapid strokes. Because water can be thrown only a short distance by this under-the-armpit method, it is used only for irrigating plot edges adjacent to canals.

Standing in or alongside a slightly widened spot in a canal or drain, irrigators scoop water out onto fields in a series of overlapping fan patterns with the irrigators at the apices (Figure 9-16). Surface moisture conditions are checked continuously, with more or less water applied to a particular part of the field as needed to achieve light puddling. Because an experienced irrigator can apply a 1- to 2-liter scoop of water to any square meter within range, application is accurately controlled. On large fields two or three men may work simultaneously on either the same or opposite sides. Young irrigators do not always ignore opportunities for a "misplaced" scoop of water or even short water fights.

The Zunil conical *pala* is worked only with a rocking motion. The handle is braced behind the shoulder or held vertically in front of the farmer, with one hand acting as a fulcrum while the other levers the scoop. Unlike the Almolonga and La Ciénaga shovel, which can lift water from canal surfaces as much as a meter below field levels, the Zunil scoop is effective only where water is near the surface. Farmers stand

Figure 9-15. Discharging water shovel over plots, La Ciénaga, Quezaltenango, Guatemala.

beside or astride the tiny, shallow canals to scoop-splash water over *tablones* that for the most part are less than *cuerda* size (Figure 9-17).

Labor Requirements

The calculated brimful capacity of the Almolonga water shovel is about 2.0 liters, and a surprising 1.7 liters is discharged on each stroke.[44] A sample of irrigators recorded at various times of day averaged 35 strokes/min, which includes both methods of scooping, brief rests, and position changes. Thus, the Almolonga and La Ciénaga irrigators achieve a delivery rate of about 60 lpm. On the average, an irrigator spends about one hour applying more than 2,000 scoops to one *cuerda* (440 m^2) of land. At this rate, he applies about 8 mm of water at each irrigation. Frequency of irrigation is the main response to increased water needs; farmers irrigate anywhere from once or even twice a day to two or three times a week, depending on soil moisture conditions.

The advantages of adequate water at or near the surface are apparent. Almolonga *pala* irrigators can pump water nearly four times

[44]I tested the several shovels, scoops, and basins in a controlled situation where water could be scooped, tossed, recovered, and measured. Reported delivery rates, therefore, may not be exactly equal to those obtained by experienced irrigators under actual field conditions.

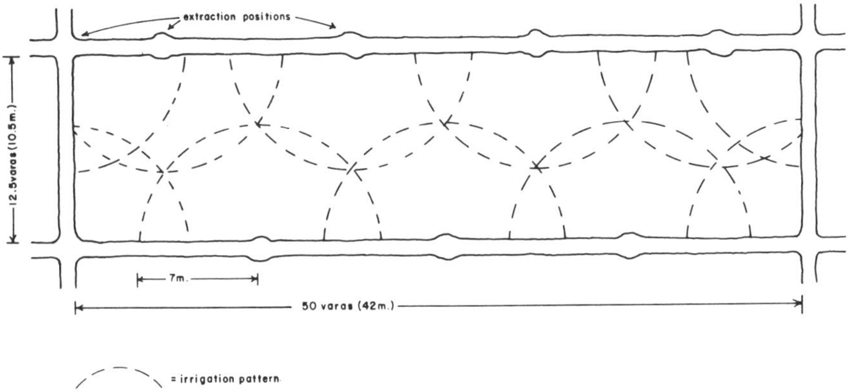

Figure 9-16. *Pala*-irrigated field, Almolonga, Guatemala.

Figure 9-17. Discharging water from tiny rock-lined canals with a conical metal scoop, Zunil, Quezaltenango, Guatemala.

as fast as Oaxacan *cántaro* farmers, who must lift and carry water from wells to fields. However, all parts of scoop-irrigated plots are wetted, so that the effective irrigated area is not as great. If we assume again that only 75 percent of working time is spent irrigating, an Almolongan farmer can irrigate 2,700 m^2, or about six *cuerdas,* a day.[45]

Calculated brimful capacity for the Zunil scoop is 1.3 liters, considerably less than for the Almolonga and La Ciénaga shovels, and the scoops deliver only a liter with each stroke. But they are levered at 50 strokes/min and thus achieve a delivery rate of 50 lpm, only slightly less than the Almolongan rate.

A few farmers are experimenting with pumps and spray nozzles to irrigate *tablones* at Almolonga. But it is doubtful that many *dueños* hold enough land to justify the investment in pumps and hoses, and pump-operating expenses may be greater than manual-irrigation labor costs.

Scoop Irrigation with Basins

In the department of Sololá, in western Guatemala, water is simply scooped out of shallow ditches and tossed over fields with metal or plastic basins *(palanganas).* Where water is at or near the surface, this is a rapid and efficient method.

Sololá is mostly mountainous, even precipitous. Extensive areas around the department capital (Sololá City) and Lake Atitlán have been carved into narrow terraces *(tablones)* on which truck (vegetable raised for market) is grown. Irrigation water for terraced hillside plots is led down in tiny canals so that each *tablón* has a water source along the uphill side (Figure 9-18). (Temperature, precipitation, and evapotranspiration data are given in Table 9-3.)

Basins are medium-size bowls 18 to 22 cm in diameter at the lip and 6 to 7 cm deep, of the type ordinarily used for food preparation and washing (Figure 9-19). They are made of lightweight steel or aluminum, or in some cases plastic, although the rigid metal basins are preferred.

To irrigate, the farmer simply grasps the edge of the basin with one hand and, resting the other on his knee, scoop-splashes water over plot surfaces. Hands are switched for relaxation, or both hands may hold the basin, especially when tossing water to the limit of the plot (Figure 9-20). Water is tossed in a sweeping vertical spray until light puddling occurs, usually after 7 to 8 mm of water have been applied. As in all

[45]One study suggests that the normal working day in Almolonga is 7 hr. T. David Johnson, "Analysis de Actividades Necesarias para la Producción de 11 Especies Horticolas (Hortalizas)," para. 4.2. However, there is great variation. I have regularized the data to an 8-hr day, 75 percent of which is spent irrigating.

Figure 9-18. Steeply terraced, basin-irrigated *tablones*, Sololá, Guatemala.

Figure 9-19. *Palanganas* used as irrigation scoops, Sololá, Guatemala.

Figure 9-20. Irrigating terraced *tablones* with a basin, Sololá, Guatemala. Terraces are about 1.5 m high.

manual-irrigation systems, farmers irrigate as needed, usually every second or third day.

To calculate labor requirements, three basins were tested, ranging in brimful capacity from 1.25 to 2.0 liters. Regardless of the size of the basin, only about one liter of water can be scooped out with each stroke. Larger charges interrupt the smooth scoop-toss motion, lessen accuracy, and produce heavy splashes. Farmers using the short, direct, arm-and-wrist motion of basin irrigation average 60 to 65 strikes/min, including short rests and moves to new locations. Thus, the water-delivery rate of 60 to 65 lpm compares favorably with the rate in the *pala* systems of Quezaltenango.

With the incentive offered by growing markets in Guatemala City and other large urban centers, the Cakchiquel farmers of Sololá are converting hillsides formerly in *temporal* subsistence production to year-round, irrigated truck. Slopes are first furrow-terraced to form planting surfaces, in some cases only 50 cm wide, which are basin-splashed from tiny canals on the up-slope side. Despite the labor involved and the small size of operations, scoop-irrigated vegetable production seems to be rewarding. Several farmers mentioned that they had abandoned their traditional migration to coastal plantations to work as laborers during the dry season[46] and were now tending their own *temporal* and irrigated plots year-round. Thus, attractive markets and a technology that can cope with rugged terrain and produce high-quality products may be changing an ancient pattern of work and migration in western Guatemala.

Scoop Irrigation with Gourds

At Aguacatán, 26 km east of Huehuetenango City, garlic and other crops are manually irrigated much as they are in Sololá, except that locally grown gourds *(huacales)* as well as metal and plastic basins are used as scoops. Fields are larger than at either Quezaltenango or Sololá, averaging about 0.3 ha, and are not as intensively worked.[47] (Temperature, precipitation, and evapotranspiration data are given in Table 9-3.) *Temporal* maize alternates with irrigated cash crops, especially the excellent garlic, which enjoys brisk national and even international demand. Fields are first prepared with ox-drawn plows,[48] then divided by hand with *azadones* into planting beds *(surcos)* 150 cm wide bordered on each side by furrows 40 to 50 cm wide and 30 to 40 cm deep. *Surco* lengths depend on field configuration and vary from a few to more than 100 m. As a final step, furrow sides may be patted smooth with wooden paddles *(paletas)*.[49] Water from small canals flowing in the furrows is checked and ponded every 2 to 3 m by small dirt-and-debris dams. Like Sololá basins and Zunil scoops, Aguacatán *huacales* need at least 5 cm water to operate efficiently.

The gourds (*Lagenaria* spp.) grow on vines in local maize fields *(milpas)* and dooryard gardens.[50] When dry, they are cut in half and scraped clean to form workable scoops (Figure 9-21). They are less comfortable to handle than metal or plastic basins, however, and their irregular configuration makes it more difficult to scoop smoothly

[46]Lester Schmid, *El Papel de la Mano de Obra Migratoria en el Desarrollo Económico de Guatemala;* Sol Tax and Robert Hinshaw, "The Maya of the Midwestern Highlands."

[47]McBryde (*Cultural and Historical Geography,* p. 32) objects to the term *tablones* for fields in Aguacatán.

[48]Smaller plots and dooryard gardens are worked with only hand tools, including broad hoes and a hand plow with an oval blade that is pulled through the soil.

[49]Smoothing furrow sides reduces infiltration. Habte Mariam Neghassi, "Effect of Smoothing on Furrow Irrigation." This procedure is desirable here because the furrows are for water delivery and not for infiltration.

[50]Donald W. Lathrap ("Our Father the Cayman, Our Mother the Gourd") argues that the bottle gourd came to eastern Brazil from Africa and, in diffusing through Middle and South America, was instrumental in the spread of agriculture.

Figure 9-21. *Huacales* used as water scoops, Aguacatán, Huehuetenango, Guatemala.

Figure 9-22. Scoop-irrigating with *huacales*, Aguacatán, Huehuetenango, Guatemala.

through the water. Irrigators stand in the furrows and, holding *huacales* with either or both hands, scoop and toss water over their long garlic beds (Figure 9-22).

Other Scoop-Irrigation Areas

Occasional scoop irrigation is so widespread in traditional Middle American agriculture that it must rank almost as a universal technique. Even in areas where flood or furrow irrigation are the usual modes, seed beds or newly planted plots are often watered by scoop-splashing. Water is taken from canals, drains, or pools with almost any convenient device, including bowls, basins, trays *(bateas),* and paddles, many of which are simply borrowed temporarily from kitchens. Like containers, scoops may be used to supply furrow-irrigation systems. For example, West reports that Tarascan farmers on the shores of Lake Patzcuaro, Michoacan, Mexico, scoop water from shallow lakeside ditches or wells into canals with a wooden, spoonlike device fixed to a long pole (*t'aparatarakua,* or *pala*).[51]

Summary

A continuous supply of surface water allows farmers to decide when and how much to irrigate. But increased control of supplies may also increase technical and organizational demands. Even simple check dams and reservoirs are subject to erosion and deposition. An elaborate distribution system and an equally elaborate allocation system are needed to manage the generous springs of Tehuacan and most river diversion systems.

In contrast, small shallow wells and suitable containers are all that are needed to tap the water table. But water from below the field level is lifted and transferred at enormous cost in human labor. The prospect of battling a relentless, evaporative sun with implements no more formidable than hand-held pots is daunting even to the compulsively industrious farmer.

Scoop irrigation may be a reasonable compromise. Lifting and applying water by shovel or basin is not as arduous as applying it by container, yet the control of irrigation amounts and timing is almost as good for the former method. Although some cooperation is required to maintain and operate the supply canals, scoop systems seem to not have the complex problems associated with surface irrigation systems.

[51]Robert C. West, *Cultural Geography of the Modern Tarascan Area,* p. 47.

Perhaps the simplicity, flexibility, and control as well as the acceptable work levels account for the persistence of scoop systems, even where these systems could be converted to gravity furrow or flood gravity systems.

Manual irrigation has other positive aspects. Water is applied gently, directly, and in proper amounts on each square meter of plot surface. Container or scoop sprinkling works well on fine-textured soils with low intake rates as well as on coarser soils if these have been modified to improve water retention, such as the heavily mulched soils of Almolonga and the *chinampas*. Timing is precise: farmers can irrigate on the day or even during the portion of the day they decide is best.[52] Manual irrigators rightly boast that their crop quality is high[53] and that plant losses due to over- or underirrigating are extraordinarily low.

[52]In vegetable production, irrigation water should be "on call." Water "on turn" is not adequate. D. W. Thorne and H. B. Peterson, *Irrigated Soils*, p. 251.

[53]The villages have reputations for product quality. Distinctive costumes worn by Indian vendors serve as "brand" identification for vegetable shoppers in Quezaltenango. Manning Nash, *Primitive and Peasant Economic Systems*, pp. 67–68.

10

Water Management of Subsurface Systems

Lifting the enormous quantities of water required even for small-scale irrigation taxes the limited energy resources of traditional farmers, especially in Middle America where there is little tradition of wind-, water-, or animal-powered lifting devices.[1] Most irrigation is gravity driven, and human effort is limited to moving material to control water flows.[2]

Within this context, ground water would seem an unlikely resource to come under traditional management. Access to subsurface water is blocked by overlying layers of earth, and there is seemingly little opportunity for gravity transfers. But water below the surface has several characteristics that partially offset these obstacles. Subsurface water is already in place, with no need for further collection and storage.

[1]Mechanical water-lifting devices are not absent in Middle America by any means. For example, *norias* (chains of buckets) and Persian wheels were transferred to the New World and survive in some areas. But the windmill landscapes of Europe (the polders in Holland, the Murcia region in Spain) and the many animal-driven wheels that are common features in rural North Africa and Spain are missing. Some of these devices have found greater acceptance elsewhere in the Western Hemisphere; see, for example, Volunteers in Technical Assistance, "The Amazing Windmills of Miramar."

[2]Not that surface irrigation is an insignificant task. John K. Hatch (*The Corn Farmers of Motupe,* pp. 61–138) estimates that more than 10 days of hard labor per season are required to flood-irrigate and maintain the irrigation facilities on a hectare of maize land.

Subsurface water is conveniently located in that it is usually coextensive with field surfaces, which minimizes the need for horizontal transfers. Even vertical transfers may be achieved by capillary movement. Thus it is that in Mexico and Central America, especially in the mountain basins where drainage is impeded and water tables are high, ground water has been exploited intensively for hundreds, in some cases thousands, of years. The techniques used are among the most sophisticated in traditional water management.

Well-and-Container Irrigation

The most direct approach to using subsurface water is simply to open wells and extract the water with containers as discussed in Chapter 9. Investments in labor and equipment are low because only one or two men with simple tools are needed to excavate the wells and operate the system. Manual well irrigation is limited to those areas where water tables are near the surface. Deeper sources, even where water is accessible, are used for domestic supplies but are too expensive for agriculture. For example, on the dry limestone peninsula of Yucatán, *cenotes* (sink holes) form natural wells, but most are too deep to supply water for irrigation. Collapsed *cenotes* are cultivated with water from runoff augmented by subirrigation.[3]

Well-and-container systems tend to be highly individualistic and do not link communities or larger social units. As Kirkby remarks:

> In contrast to canal irrigation, the amount of water available to plants in a pot [container] irrigation from wells system is dependent upon the decisions of individual irrigators. Thus, within the limits imposed by his labour, he [the farmer] can irrigate as many plants as often as he likes, independent of community control.[4]

Well-and-Pump Irrigation

Small gasoline or electric pumps are an obvious alternative to arduous manual container-and-well irrigation. Frequently they are the first pieces of powered equipment to appear on traditional farms (Figure 10-1). But even small pumps require substantial initial investments and continuing outlays for fuel and parts. Furthermore, well flows and farm

[3]George Cheever Shattuck, *The Peninsula of Yucatan,* p. 23.

[4]Anne V. T. Kirkby, *The Use of Land and Water Resources in the Past and Present Valley of Oaxaca, Mexico,* pp. 113–14.

Figure 10-1. Small pump extracting irrigation water from hand-dug well.

sizes must be large enough to supply and absorb pump discharges for the operation to be economical. Joint or cooperative ownership offers the possibilities of shared costs and economies of scale. But it also presents problems of maintenance and scheduling, which farmers accustomed to the simplicity and independence of container irrigation find difficult to overcome.[5]

The situation is quite different around Tepeaca, Mexico, where deep aquifers and a long tradition of cooperative water management are found. Tepeaca lies less than 100 km northwest of the Tehuacán Valley and shares many of its hydrological and cultural characteristics. Among them is the cooperative irrigation society *(sociedad de agua)* for managing spring water. As deep-well technology invades the region, the *sociedad* organization is being transferred to well-and-pump operations almost unchanged and with notable success.

[5]Several Mexican government programs not only help finance and install small, 2- to 4-in. pumps but also organize five to ten farmers to cooperate in their use. Other programs offer similar assistance with deep-well pumps. But because these have discharges on the order of 80 lps, they involve considerably larger groups of farmers. Daniel W. Bromley (*Improving Irrigated Agriculture*) discusses farmer independence in four situations: individually operated pumps and individual water distribution, and jointly operated pumps and joint water distribution.

Sociedades handle both start-up and running costs on the basis of shares. Drilling costs are substantial as are those of pumps—which vary in size from 4 to 10 in. (10 to 25 cm), although 6 in. (15 cm) seems most popular—and related equipment. Total investments run as high as US$20,000. Members *(socios)* put up capital for well perforation and equipment, with shares allotted on the basis of contribution. Once pumps start to produce, associations assess hourly charges to cover costs of operation and maintenance. There usually are fifteen shares *(acciones)* of 24 hr twice a month, which coincides with the normal period of irrigation in this region. Shares can be divided in halves (12 hr) or quarters (6 hr), or a *socio* may own more than one *acción*. In the Tepeaca region well-and-pump societies have from ten to thirty or more members.

Although technical and related problems of deep-well drilling are substantial and capital requirements are formidable for traditional farmers, the *sociedades* generally eschew government offers of technical and financial assistance, claiming that government participation results in lower efficiency, higher costs, and loss of autonomy.[6] If we disregard the ethical question of whether such groups should continue to control a public good, the Tepeaca *sociedades* is an outstanding example of a traditional organization adapting to a radical innovation (deep-well, powered pumps) and, in the process, exposing its members to the problems and decision making associated with modern technological and financial management.[7]

Galerías

The *galerías filtrantes*[8] of the state of Puebla, Mexico, offer quite a different solution to the problem of ground-water extraction. Essentially *galerías* are tunnels sloped more gently than natural gradients so that eventually they intersect the surface (Figure 10-2). Water is thus free to

[6]Middle American traditional farmers retain an abiding suspicion of government agents and activities and a fear of bureaucratic entanglements. The attitude is common among small-scale farmers everywhere and, often as not, is well justified.

[7]The closely related water-management organizations of the Tehuacán Valley are reviewed by Scott Whiteford and Luis Emilio Henao, "Irrigación Descentralizado, Desarrollo y Cambio Social."

[8]Also called horizontal wells, or chain wells after the line of access shafts that mark *galería* alignments on the surface. In the Old World they are known as *qanats* (Iran), an ancient Semitic word and ancestor of the word *canal; karez* (Afghanistan and Pakistan); *foggaras* (North Africa); and *fajal* (United Arab Emirates)—all of which have alternate English spellings. National Academy of Sciences, *More Water for Arid Lands,* p. 56

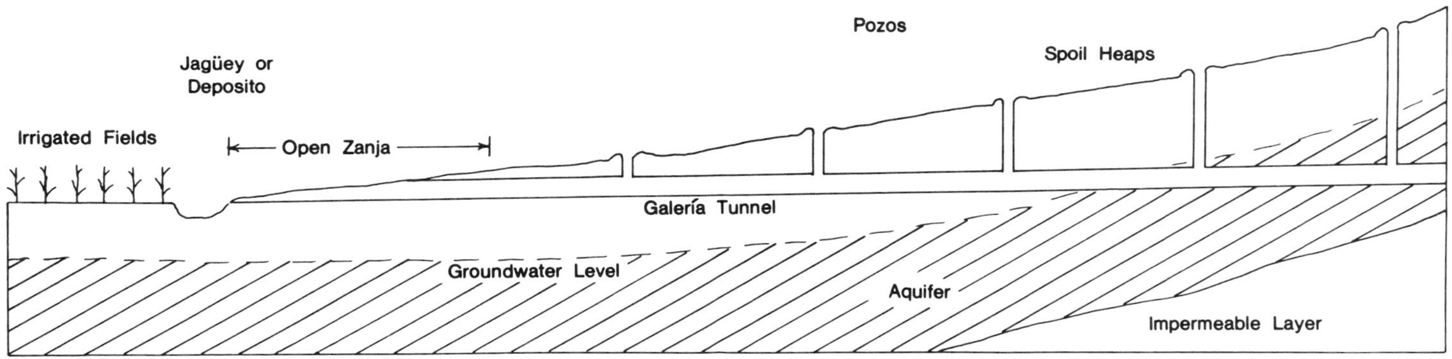

Figure 10-2. Cross-section of lower portion of *galería filtrante*.

Table 10-1. Partial List of Chilac *Galerías*

Galería[a]	Flows (lps)[b]
Agrícola, or San Miguel	9
La Ascensión	10
La Aurora	56
Chilac-Tlacoxcalco-Tetitlán (a combined *galería* system)	93
La Guadalupana	4
Mercedes de Maria	40
La Purisima	5
San Gabriel	23
San Isidro	7
Total	247

SOURCE: Files of the Secretaría de Recursos Hidráulicos, Tehuacán.

[a]There may be twice this number of *galerías* in the area, some no longer flowing, some combined with those listed, and some simply unreported.
[b]Flows recorded 1973–1974.

flow by gravity from subterranean sources. *Galería* tunnels are 80 to 100 cm wide, 150 to 180 cm high, and anywhere from less than 100 m to several kilometers long. Vertical shafts *(pozos, lumbreras)* up to 40 m deep spaced every 30 to 60 m connect the tunnel to the surface.[9] *Pozos,* originally opened to allow removal of excavated tunnel material, continue to serve as air vents and for access to the tunnels when maintenance is necessary. At the source end *galerías* penetrate water-bearing strata or alluvium. At the discharge end they appear first as deep open channels *(zanjas),* which gradually become shallower until the water enters small reservoirs *(jagüeys, depositos)* or flows directly into canal-irrigation systems. *Galería* flows range from a few to a hundred or more liters per second (Table 10-1).

Although *galería* waters are "filtered," as are all subsurface waters, they are not necessarily of high quality. In Tehuacán, *galería* and spring water ranges from nearly pure to delicious mineral waters (bottled by such well-known companies as Garci-Crespo, Penafiel, and

[9]*Galerías* up to 30 km long and *pozo* shafts 70 m deep are reported in the literature. Richard K. Cleek, "The Infiltration Gallery in Mexico"; Enno Seele, "Galerías Filtrantes en el Area de Acatzingo-Tepeaca, Estado de Puebla"; Enno Seele, "Galerías en el Estado de Puebla"; Whiteford and Henao, "Irrigación Descentralizado."

San Lorenzo) to flows unfit for drinking. Similarly water temperatures in *galería* tunnels range from 15° to 26°C.[10]

Galerías are possible only where certain geological and hydrological conditions are favorable. Slope is critical: if too gentle, tunnels could never reach the surface; if too steep, ground water would be inaccessible. Water tables must also be located appropriately and supplied from effective recharge areas. The travertine escarpments and alluvial fans of the state of Puebla, Mexico, have proven especially suitable for *galería* development. Seele lists more than 100 for the Tepeaca-Acatzingo region and estimates that the total may exceed 150. The Tehuacán Valley probably has as many if not more.[11] Not all are still flowing. Around Tepeaca and Acatzingo water tables are falling, primarily because of modern deep-well pumping, and more than half the *galerías* no longer carry water. However, in the Tehuacán Valley surface irrigation from the huge Valsequillo dam and canal system is raising water tables and increasing *galería* flows.

In 1958, when Cressey reviewed *qanats,* primarily in the Old World, little was known about their New World counterparts.[12] Subsequently the *galerías filtrantes* of Mexico have enjoyed considerable attention. In addition to the major concentrations in the Tepeaca-Acatzingo and Tehuacán regions of Puebla, *galerías* are also reported in the states of Coahuila, Colima, Guanajuato, Queretaro, Tlaxcala, and Zacatecas.[13]

Management

As we have seen, traditional subsurface water systems are generally not conducive to the formation of irrigation societies. Subsurface water

[10]Seele, "Galerías Filtrantes." For chemical analyses of Tehuacán water see Luis Blásquez L., "Hidrogeología de una Parte de Tehuacán, Edo. de Puebla"; Borja León Marquez, "Breve Estudio de las Condiciones Sociológicas e Higiénicas del Municipio de Sn. Gabriel Chilac, Tehuacán, Pue." Temperatures are from Blásquez L. and field measurements.

[11]Seele, "Galerías Filtrantes"; Seele, "Galerías en el Estado de Puebla." Whiteford and Henao, "Irrigación Descentralizado."

[12]George B. Cressey, "Qanats, Karez, and Foggaras." For a summary of *qanats* in western Asia and northern Africa see Paul W. English, "The Origin and Spread of Qanats in the Old World." Carl Troll ("Qanat-Bewässerung in der Alten und Neuen Welt") first called attention to the similarities between Tehuacán *galerías* and Old World *qanats.* Just how this water technology was transferred to the New World or whether it had pre-European precedents is not known. Seele, "Galerías Filtrantes"; Seele, "Galerías en el Estado de Puebla"; Richard B. Woodbury, "Prehistoric Water Management Systems in the Tehuacán Valley, Mexico."

[13]Cleek, "The Infiltration Gallery in Mexico"; Seele, "Galerías Filtrantes"; Seele, "Galerías en el Estado de Puebla"; Whiteford and Henao, "Irrigación Descentralizado."

is not located in only one place as are channeled surface flows such as streams or canals. Phreatic (water-table) surfaces are commonly ubiquitous within a small area and farmers have equal access via inexpensive and independently dug wells or open drainage ditches. They have such access, however, only when water tables are shallow. Deep water is inaccessible to traditional farmers not so much because of the technical difficulties of digging wells as because of the prohibitively high labor cost of manually extracting irrigation water from depths much below 10 m.[14] Furthermore, ground water requires vertical rather than horizontal transfers; horizontal transfers make storage, distribution, and allocation possible with surface flows and call forth attendant organization and management. However, *galerías,* although they tap deep water not readily available to individual farmers or farm families, are based on a completely different concept of ground-water extraction than are vertical wells. The long, gently sloped *galería* tunnels convert relatively static ground water to surface flows.

But *galerías* are costly to build and maintain. Although only simple hand tools are used, most of the construction must be done by specialist well diggers *(pozeros).* Farmers can contribute only a minor share of the required investment as labor; the rest must be paid in cash. Such an investment is far too high for most traditional small-scale farmers.

The solution is to form organizations that are essentially partnerships. Before constructing a new *galería* twenty to sixty farmers form a *sociedad de agua* under a name that will apply to both the association and the *galería.*[15] Membership is based on friendship and family ties and on location, although some members may be speculative investors and not hold land in the area to be served by the *galería* or may not even be farmers. In addition, prospective members must be financially sound, able to contribute to original construction costs and subsequent maintenance and possible expansion.

Estimates of construction costs determine initial capital requirements and thus the cost of each *acción.* Some surplus is advisable to cover additional expenses such as those engendered by excavation through unexpected rock. But the ultimate flexible factor is time; like other traditional projects *galerías* are not constructed on rigid schedules or with rigid deadlines. Excavation may take years, and progress often depends on the rate at which *socios* can set aside construction costs from limited operating budgets.

[14]The higher value and smaller volume of potable water than of irrigation water allow manual extraction from much deeper sources. One Maya *cenote* source was found 137 m below the surface. J. L. Stephens, *Incidents of Travel in Yucatan,* as reported by Ray T. Matheny, "Maya Lowland Hydraulic Systems."

[15]Like springs, *galerías* have Spanish, Christian, or Nahuatl names.

Usually a new *sociedad* issues sixty *acciones*, each of which entitles the holder to 12 hr of *galería* flow per month. *Socios* participate according to the amount of their investment, from a fraction to several *acciones*, although usually not more than four or five are in the hands of any one member.[16] If at some future time a member decides to sell his shares, he must give other *socios* first right of refusal before offering them to nonmembers of the association.[17] Although *galería* operating costs are low, periodic cleanings and routine upkeep are needed. In some associations members contribute cash toward maintenance costs; in others they contribute labor. When labor is contributed, all *socios* are expected to participate; those who shirk are fined.

Sociedades must obtain permission to construct *galerías* from villages and individuals along the proposed route. Although the tunnels themselves do not disrupt the fields beneath which they pass, access shafts and spoil heaps become permanent features on the surface. Also, the question of whether a new *galería* will alter local hydrology and flows of existing water systems receives considerable attention. Some villages charge for access rights, a fee that becomes part of the initial capitalization.

Day-to-day operations of *galerías* are managed by a president, secretary, treasurer, two *vocales* (literally, spokesmen) or general representatives, and perhaps other minor officials. Offices are rotated each year, and all *socios* are expected to take their turns irrespective of number of shares held. In addition, some associations have permanent *gerentes*, wise and respected members who act as mediators of disputes between *socios* and as intermediaries between the *sociedad* and other organizations. Associations may employ *canaleros*, or canal tenders, to regulate distribution of *galería* water down to the point of division to individual farms, bookkeepers to keep accurate records, and, of course, the specialized *pozeros* to do actual construction and maintenance.

In dry western Puebla irrigation once a month may not be adequate, and 12 hr of flow from a poorly producing *galería* would not suffice. Exchanges and sales of *galería* water help adjust supplies and demands. Usually trading is between neighboring farmers. However, in San

[16]Seele ("Galerías en el Estado de Puebla") says each share is for 24 hr. My own field inquiries indicate that the 12-hr *acción*, also reported by Whiteford and Henao ("Irrigación Descentralizado"), is more common. Although most *sociedades* have fractional shares of only halves (6 hr) or quarters (3 hr), a *galería* near Zinacatepec is reported to have 175 *socios*, some of whom have but one-twelfth of a share, or one hour of flow. In some cases two or three farmers jointly purchase an *acción*, thus effectively creating further division.

[17]Financial details of *sociedades de agua* are difficult to learn or confirm. I have heard of *acciones* valued at the equivalent of US$7,000. Although this investment is enormous for small-scale farmers, it is quite reasonable considering the value of the water it represents.

Gabriel Chilac an active water market (described in Chapter 9) convenes daily for sales of water turns or rentals of seasonal rights. In other communities sharecropping with farmers who have land but no water is common. Landowners provide land, labor, seed, and fertilizer; *acción* holders provide water. Profits are split equally.[18] Thus, although initial risks are high, there are opportunities for substantial profits in addition to the agricultural benefits of an assured water supply. In these ways *galería sociedades* resemble capitalistic institutions more than they do the socialistic organizations commonly associated with surface-irrigation systems.

As can be seen from this description, *galería sociedades* differ in several respects from surface-irrigation societies. *Galería sociedades* involve initial capital contributions from a small group of investors. The chances of a dry perforation, and thus complete loss, are high. But if the venture is successful, the water belongs to the *sociedad* in a proprietory sense quite contrary to the ancient and modern Mexican concept of using resources for the communal good. Shares in *galerías* with abundant flows increase enormously in value.

The concept of water as private property is not new in Tehuacán. Whiteford and Henao suggest that the tradition has its roots in colonial or even precolonial times:

> The ideology of economic equality, reportedly characteristic of many Indian and peasant communities in Mexico, has long been replaced in the Tehuacán Valley, if it ever existed, by a capitalistic ideology of hard work and individual acquisition. It is possible this is a product of a series of factors associated with irrigation: a precolumbian stratified system within the larger communities which was perpetuated by a class of indigenous chiefs *(caciques);* early incorporation of the peasantry as wage laborers on the haciendas which gained control of community land and water rights; and the increased involvement in commercial agriculture by those who had water. [Within this context] the irrigation associations represent a combination of pre-existing traditional elements of the social organization adapted to a new enterprise.[19]

In the examples in Chapter 9, shareholders of irrigation associations formed around spring-supplied water enjoyed such control over specified periods of flow *(acciones)* that they could freely use, trade, and even sell their *acciones.*[20] *Galería socios* go an important step further. Their

[18]Whiteford and Henao, "Irrigación Descentralizado," pp. 64–65.

[19]Whiteford and Henao, "Irrigación Descentralizado," p. 62.

[20]Scarce and often expensive spring and *galería* water is delivered with considerable punctuality, even to the point of allowing for travel time in the canals between farms. Turns of water are cut off two or three minutes before the specified time so that successive recipients do not lose cumulative water-delivery time. The amount deducted is called *la discuenta* (the discount).

capitalistic perspective arises perhaps from the fact that, unlike free-flowing springs, the deep ground water tapped by *galerías* is otherwise inaccessible. *Galería socios* believe they have created an economic good (irrigation water) out of a resource unusable in its natural state (ground water) with expensive technology *(galerías)*, and they feel that the investments and risks merit whatever rewards ensue.

But because the *sociedad* approach is in conflict with the concept of water as a public or common good, the Tehuacán Valley has become a sometimes tense testing ground where government, in the form of the Secretaría de Agricultura y Recursos Hidráulicos and the Paploápan River Commission, is attempting to extend control over subsurface waters. The *sociedades* continue to operate as private enterprises. But no new *galerías* may be built without government permission, and even maintenance of existing systems, which often includes tunnel extension, requires official approval. Not surprisingly, the *socios* steadfastly resist efforts to diminish their control.

Construction and Maintenance

Tunneling poses its own set of distinctive and difficult problems, especially for traditional societies without powered equipment or modern surveying methods. Tunnels are run through everything from loose and unstable sand to solid rock and are aligned accurately with only occasional access to surface reference points. *Galería* gradients and flows must be precisely controlled to avoid either ponding or erosion. Physical movement in the narrow tunnels is restricted and awkward. Only the promise of substantial rewards or the press of dire necessity will induce extensive tunneling by traditional methods.

Locating Water

The initial objective is to minimize the risks of a nonproductive *galería*. One or several methods may be used; perhaps the most common and most successful is to excavate in an area where wells, springs, and other *galerías* attest to the presence of water.[21] Perhaps a water diviner *(visionario)*, usually someone with long experience in the region, will look for topographical, geological, and botanical clues (for example,

[21]This approach can cause conflict if owners of existing *galerías* fear that the new installation will tap their sources and reduce flows. Their fears may be well founded. In some reported cases new perforations were made close to or even under other tunnels or springs in a deliberate effort to capture water that was already being exploited. Such conflict has sometimes resulted in violence. Scott Whiteford and Luis Emilio Henao, "Commercial Agriculture, Irrigation Control, and Selective Labor Migration."

water-indicating grasses and trees) to identify likely spots for a new well. Still, risks remain high. Whiteford and Henao report that 50 percent of the chain-well projects in the *municipio* (roughly, county) of Ajalpan fail and suggest that this figure may hold for the rest of the Tehuacán Valley.[22]

Tools and Procedures

Once association leaders and technical advisers have agreed on a general course and secured necessary approvals from the government and villages along the way, digging can commence. When *galerías* are first laid out, the location of each *pozo* is marked. A cord establishes straight-line tracks between wells. As the tunnel is excavated away from a *pozo,* two weighted cords perfectly aligned with the surface track are dropped down the shaft, often with a candle behind. If *pozeros* can see both cords, they know that the tunnel is out of alignment. *Pozeros* use simple but effective methods to maintain tunnel gradient and alignment. In the past they determined gentle slopes on the order of 1:1,000 by pouring water on tunnel floors. Now, spirit levels are common.

Tools used for construction are few and simple: picks, shovels, spades, and wheelbarrows for general excavation and materials handling, a heavy (3- to 4-kg) hammer and rock chisels and spikes *(puntas)* for dislodging consolidated material, a wooden windlass, rope, and special woven palm-leaf baskets (*chatenates,* or often the shortened forms *tenates* or *tanates,* from the Nahuatl *tanatli*) for lifting excavated materials, and carbide lamps, candles, and cord for illumination and surveying. Blasting powder and fuses are added to the tool kit if the tunnel must pass through substantial amounts or rock, as are special steel or light concrete *(tabique)* reinforcing frames *(broncales)* if sand or other loose materials are expected.

Only two or three *pozeros* work in the tunnel at a time: one or two at the face dislodging material, and another loading *chatenates* for the lift to the surface. One or two workers, often young apprentices,[23] remain at the shaft head to man the windlass and dispose of excavated material. Thus, a complete *galería* construction crew consists of but three to five workers.

Except in portions of tunnels near outlets, all personnel, tools, and excavated materials are transfered to and from the surface through access shafts. The hand-excavated *pozo* shafts are usually oblong, 60 to 100 cm wide, 120 to 150 cm long, and up to 40 m deep. Shaft openings may be reinforced with logs to keep edges from collapsing. As soon as well

[22] Whiteford and Henao, "Commercial Agriculture," p. 5.

[23] Although sons often follow their fathers into *pozero* work, entrance to the craft is through apprenticeship.

Figure 10-3. *Pozero* descending *galería* shaft in rope sling. Note log windlass supported by forked sticks and turned with branch handle.

diggers can no longer shovel debris directly from deepening *pozo* shafts, they erect simple windlasses *(turnos)* with which to hoist excavated material to the surface. Two forked sticks *(horcones)* embedded on either side of a shaft opening support a smoothed log 10 to 15 cm in diameter; the log is perhaps 2 m long over reinforced shaft openings but a meter or so longer over unreinforced *pozos,* where the forked-branch supports must stand well back from the opening. A smaller branch *(orqueta, manilla)* serves as a handle and provides a mechanical advantage. Wood from local *pirú* (or the Nahuatl *huavino*) trees *(Schinus molle)* is preferred. Tough mesquite (*Prosopis* spp.) branches serve if suitable *pirú* is not available.

The sturdy branches and mechanical advantage are needed. The woven palm *chatantes* hold up to 0.1 m^3 (about 100 kg) of material. Workers descend and ascend *pozo* shafts sitting in a simple loop sling (Figure 10-3). *Pozeros* claim there is little danger in the tunnels themselves but mention accidents when ropes have broken during trips up or down the shafts.

The tunnel itself is pitch black, cool but not cold, and humid. Near the working face the tunnel is littered with the debris of excavation. Dislodged earth and rock are shoveled into baskets that are lugged to the nearest *pozo* and attached to the rope dangling down from the surface.

Niches along the tunnel walls hold lamps, tools, coffee jars, and underwear, which *pozeros* don when ascending to the surface. In the damp and muddy privacy of a *galería, pozeros* usually work nude. A shallow stream flows along the bottom if the tunnel has already perforated water-bearing strata.

Excavation Rates and Labor Costs

Excavation progress depends on three factors: available capital, tunnel depth, and subsurface materials. I mentioned earlier that construction might be delayed or slowed by shortages of money to pay wages. Although perforating a deep *galería* poses no special technical problems, the time required to excavate access shafts and subsequently transfer men, tools, and materials is greater for deep than for shallow tunnels.

But the primary determinant of construction progress and costs is the material that must be removed. Unfortunately *pozeros* do not keep precise records of excavation rates. Furthermore, other variables such as tunnel depth and crew efficiency complicate the calculation. Nevertheless, from several Tehuacán *pozeros* and *sociedad* presidents I have corroborating estimates that provide a rough range of excavation rates and costs (Table 10-2).

Usually *sociedades* contract with *pozero* crews for excavation at an agreed rate per linear meter or per week. Alternatively, *pozeros* may work *por estajo* (from *destajo,* piece of work) at a fixed fee for a given task. Combining all estimates I determined a cost range (at the 1974–1975 exchange rate) of US$50 to $80/week (6 days of 6 to 7 hr each). At these rates *galería* tunnels cost from US$4 to $80/m (linear), or roughly US$2.85 to $57.15/m^3 to excavate. These figures do not include costs of sinking access shafts, or of reinforcing loose sand, or where *galerías* go under roadways.

The range of costs for excavating different materials is quite wide and suggests the difficulty *sociedades* face in establishing initial capital requirements. Although by now the dozens of *galerías* that crisscross the Tehuacán Valley have revealed most of the major subsurface characteristics, an unanticipated rock layer can slow construction for weeks and add substantially to costs. Given the limited financial resources of most farmers, such an obstacle might cause work to cease entirely until the *socios* can raise additional funds. Similarly, if a *galería* does not produce water as anticipated, the *sociedad* must decide whether and when to undertake further exploration.

Under the most advantageous conditions (*blando* material) skilled *pozeros* could conceivably tunnel a kilometer in less than 2 yr. But such easy and sustained effort is unlikely; most *galerías* are built in fits and

Table 10-2. *Galería* Excavation Rates and Costs

Material	m/Week	m^3/Week[a]	Approximate Costs at US$50–$80/week[b] per m	per m^3
Soft soil (*blando*)	12	16.8	4.00–6.65	2.85–4.75
	9	12.6	5.35–8.90	3.80–6.35
	6	8.4	8.00–13.35	5.70–9.50
	3	4.2	16.00–26.65	11.45–19.05
	1.5	2.1	32.00–53.35	22.85–38.10
Hard rock (*duro*)	1	1.4	48.00–80.00	34.30–57.15

[a]Cubic meters based on typical *galería* tunnel 0.8 m wide × 1.75 m high = 1.4-m^2 cross-section.
[b]At 1974–1975 exchange rate. Range compiled from daily, weekly, and piece-work estimates from various field sources.

starts through a variety of materials. Long tunnels take many years to complete. There are reports of *galerías* in construction for 15, 20, or even 30 yr. Because even fully functioning *galerías* are frequently extended in search of additional water, it is difficult to say just when, if ever, one is truly finished.

A completed and operating *galería* still needs *pozero* attention for routine maintenance, tunnel repair, and shaft cleaning. Maintenance is essential; the shallow, low-gradient streams have only limited self-cleaning power and even small debris falls can block or diminish flows.[24] *Pozeros* watch constantly for seeps along tunnel walls, which reveal minor veins of water that can then be developed to add to the flow. For such routine work *pozeros* receive a flat daily wage, which in 1974–1975 was the equivalent of about US$3.

Advantages

The result of all this organization and effort may be a generous flow of water that will bring relative prosperity to the *sociedad* and its members. Or it may be an insignificant discharge, or even nothing at all. Some *galerías* are victims of falling water tables; others have been dry or nearly so since they were built.

[24]Sharp earthquakes in this seismically active region send debris tumbling down to block water flows within *galerías*.

Figure 10-4. *Galería* warning: "Danger, Zone of Wells," Tehuacán Valley, Mexico.

But despite substantial risks, it appears that potential rewards justify continued investment. In the Tehuacán Valley new *galerías* are still being constructed and old ones maintained. Because not all are officially recorded and fewer still are metered, total discharge is uncertain. One government engineer speculated that total flows of the Tehuacán *galerías* alone might amount to 8 m^3/sec. Seele suggests that the combined flows of *galerías* in the Tehuacán and Tepeaca-Acatzingo regions are comparable to the annual irrigation capacity of the huge Valsequillo dam.[25] In some areas *galerías* are so prevalent that signs warn of subterranean tunnels and open *pozos* (Figure 10-4).

The lure of a freely flowing *galería* must be viewed within a regional context. In the semiarid, frost-free Tehuacán Valley and the slightly wetter and cooler Tepeaca-Acatzingo region multiple cropping of high-value produce and flowers is possibly only with irrigation. *Galerías* are dependable, convenient sources of water. In addition, they have certain characteristics that appeal to nonindustrial farmers. Unlike power-drilled vertical wells, *galerías* can be constructed in stages, and *galerías* require only locally available technology to build and maintain. Although initial investment and risk are high, there is always

[25]Seele, "Galerías Filtrantes"; Seele, "Galerías en el Estado de Puebla."

the chance that a *galería* will be a heavy producer and greatly increase the farmers' wealth and income. Operating costs are low; unlike pumps, *galerías* need no fuel or mechanical repairs. Social organizations for building and operating *galerías* are already part of the local culture. Thus, *galerías* are an effective water-supply system, and construction and operation are completely within the technical and managerial competence of traditional farmers. Little wonder that they have persisted through the centuries and remain popular today.

The future of *galerías,* however, is uncertain, assailed as they are on one side by expanding pump technology and falling water tables and on the other by persistent government efforts to control them. But under proper hydrogeological conditions *galerías* still offer an attractive alternative to deep pumping. In fact, Mexican government engineers have found that modified versions of *galerías* serve admirably as potable water supplies for towns and villages throughout central Mexico.[26] Perhaps modernizing *galería* technology with improved surveying and excavating equipment would allow this exceptional system to continue and even expand.[27]

Subirrigated and Drained Fields

In the preceding discussions of water management, water transfers have emerged as a critical element. At this stage, traditional farmers face the vexing requirement of controlling the movement of the enormous quantities of water needed for agriculture with the limited power at their disposal. Gravity is the great asset here and is employed as the transfer agent whenever possible. An extreme example is the *galería filtrante,* which makes it possible to extract underground water by gravity. The only alternative to gravity transfers reviewed so far is human power, which is adequate for only tiny fields.

But there is another force which, although effective only over short distances, is capable of lifting all the water required for even the most demanding crop. This is capillary movement, which is caused by the adhesion of water to surrounding surfaces and the cohesion of the water molecules themselves. In a sense, subirrigation is drainage in reverse,

[26]Cleek, "The Infiltration Gallery in Mexico."

[27]James Sanchez, "Qanats as a Viable Contemporary Irrigation Technology." *Qanats (galerías)* are included on the National Academy of Sciences list of "promising research opportunities." *More Water for Arid Lands,* pp. 56-59. Cased horizontal wells are used for watering stock on the San Carlos Apache Reservation in Arizona. W. T. Welchert and Barry N. Freeman, "'Horizontal' Wells."

with upward capillary movement replacing downward, gravity-driven percolation.

Although capillary movements take place in all directions, we are concerned here primarily with movements upward from the water table, in a zone called the capillary fringe. The height of this zone depends on soil textures; it ranges from less than 30 cm in coarse sand to more than 2 m in fine clays. A range of 40 to 120 cm would probably cover most agricultural soils. The rate of movement also varies with soil texture; in coarse sands capillary water approaches maximum height in just a few days, whereas in fine clays and silts it takes much longer. The result of these factors is a zone, saturated at the bottom immediately above the water table and becoming progressively drier upward, whose height and water-transfer rates depend largely on soil texture. Despite these variations, however, water cannot move upward in most soils at a rate sufficient to supply plant needs for more than about a meter.[28]

The capillary fringe is important because few plants are able to send roots into saturated soils below the water table. Although high moisture levels are not necessarily harmful to plant growth, aeration in waterlogged soils is deficient and deeper roots will not develop or, if already grown, will die. The capillary fringe offers a favorable environment in which both air and water are available.[29]

However, three major management problems are associated with capillary irrigation, or subirrigation. First, because the height of capillary rise is limited, a supply source, in this case the water-table surface, must be within a meter or so of the root zone. This hydrological requirement is met rather extensively in Middle America.

Second, ground-water levels and thus the capillary fringe change over time and often from season to season. Perhaps more important, different plants and the same plants at different growth stages have different rooting depths and water consumption rates. Matching capillary fringe with root zones and water use becomes a management problem of some complexity.

Finally, some of the water transferred upward converts to the vapor phase in the soil or at the surface, precipitating material in solution.

[28]Nyle C. Brady, *The Nature and Properties of Soils,* pp. 178–81; Firman E. Bear, *Soils in Relation to Crop Growth,* p. 77; W. R. Gardner, "Soil Water Relations in Arid and Semiarid Conditions"; C. Kalisvaart, *Subirrigation in the Zuiderzee Polders,* pp. 17–18; C. L. Tolman, *Ground Water,* p. 148; Gardner, "Soil Water Relations."

[29]L. J. Evans, "The Responses of Crops to Poor Soil Drainage"; J. Letey, L. H. Stolzy, and G. B. Blank, "Effect of Duration and Timing of Low Oxygen Content on Shoot and Root Growth"; Rosemary J. More, "Water and Crops"; Tolman, *Ground Water,* p. 148; John E. Weaver, *Root Development of Field Crops,* pp. 63–64; J. Wesseling et al., "Land Drainage in Relation to Soils and Crops."

Ideally subirrigation systems should be capable of adjusting the relationship between plot surface and water table in order to maintain favorable moisture content in root zones and also to balance water withdrawn from the capillary fringe by evapotranspiration.[30] The extent to which water tables need to be controlled depends on soil texture. As previously explained, in sandy soils capillary rises are not high and soil moisture content varies appreciably with every change in the water table. Clayey soils have higher capillary fringes and soil moisture content varies less. In sandy soils ground-water tables should be controlled to within 10 cm. Clayey soils allow somewhat greater latitude but still should be controlled to within 30 cm.[31]

In industrialized systems ground-water levels are adjusted with powered pumps.[32] But traditional farmers lack such equipment,[33] and the alternative of adjusting plot heights involves soil transfers that are beyond traditional capacities, at least over a short period of time. The most common solution is to employ a mix of irrigation systems: furrow or manual irrigation when plants are young, and subirrigation at later stages when roots have penetrated the capillary fringe. This procedure requires mastery of several distinct irrigation procedures, a requirement that apparently does not dismay resourceful traditional farmers.

Conditions favorable to subirrigation occur primarily in two general zones in Middle America. The best known is the intermontane basins that stretch from Central Mexico southeast into Costa Rica. Late Cenozoic vulcanism blocked normal stream drainage in many of these valleys and basins and created vast lakes where alluvium was deposited. Although subsequent erosion partially restored drainage networks, high water tables and open swamps and lakes remain. In addition, heavy soils with poor internal drainage or impermeable layers of clay or *tepetate* underlie agricultural soils.[34]

The other region is the Atlantic coastal lowlands, where vast areas of land are periodically flooded and where evidence of prehistoric land

[30]Food and Agriculture Organization and United Nations Educational, Scientific, and Cultural Organization, *International Source-Book on Irrigation and Drainage of Arid Lands in Relation to Salinity and Alkalinity,* p. 430. See also Jerzy Jaworski, "Evapotranspiration of Plants and Fluctuation of the Groundwater Table."

[31]Food and Agriculture Organization and United Nations Educational, Scientific, and Cultural Organization, *International Source-Book;* Roy L. Fox, John T. Phelan, and Wayne D. Criddle, "Design of Subirrigation Systems."

[32]Subirrigation has been defined as "the act of regulating the elevation of the groundwater table by artificially adding water underground." Wayne D. Criddle and Cornelis Kalisvaart, "Subirrigation Systems."

[33]Except for traditional farmers with wind-powered devices, who can control ground-water levels.

[34]Robert C. West and John R. Augelli, *Middle America,* pp. 25–26.

reclamation has recently come to light. Because little is known about farming and irrigation methods on these ancient drained or raised fields, the discussion here will focus on subirrigation techniques in the highlands.

Irrigation Method

Rare indeed is the naturally subirrigated field. Although wet lands (*tierras de humedad* or *de jugo,* moist or juicy lands) are common, specific requirements of crops rule out their unmodified use except for less demanding pastures (wet meadows). Instead, subirrigated fields destined for farming are carefully leveled and provided with some means of maintaining water levels in relation to field levels.

In Middle America the nearly universal approach is to construct networks of open drainage canals (*zanjas*). Field drains bordering narrow plots are linked to form grids, and connected to master drains *(zanjas madres)* that discharge into perennial or intermittent streams or in some cases into swamps or other still-water bodies. The open drains are multipurpose: they limit maximum height of water tables, supply water for subirrigation and for surface irrigation when plants are young and during dry periods, and carry away surface runoff from heavy rains.[35] In addition, drains supply field dressing in the form of deposits rich in organic matter and minerals. Large drains may become transportation routes for boats carrying farmers and farm products between plots and even from fields to markets, as in the *chinampas* of the Basin of Mexico. The drains themselves contain useful plants, such as watercress and *tule,* and water fowl, fish, and smaller fresh-water organisms.[36]

Hydraulic characteristics exert a powerful control over the form of subirrigated fields. The objective is to achieve relatively rapid transfers of water from *zanjas* into plot substrates during dry seasons, and reverse flows from plots to *zanjas* during rainy periods to remove excess moisture. Flows are substantial. With an evapotranspiration rate of 3 mm/day, not unusual in Middle American latitudes, a fully vegetated plot 10 × 100 m would discharge 3 m^3 each day.[37] This water must move through the plot, in this case for an average distance of 2.5 m. The

[35]Separate irrigation and drainage networks are not common in traditional subirrigation systems. Criddle and Kalisvaart ("Subirrigation Systems") review the advantages and disadvantages of separated and combined networks.

[36]Gene C. Wilken, "The Ecology of Gathering in a Mexican Farming Region."

[37]Although actual evapotranspiration is usually lower than potential evapotranspiration even under intensive cropping, the potential rate suggests the general volume to be considered. Chaudhry Nuruddin Ahmad, *Plant Uptake of Water from a Water Table,* p. 69.

essential point is that efficient subirrigation and drainage require rapid horizontal water movement, and, regardless of soil texture, this movement is possible only across narrow plots.

Lack of systematic surveys makes it difficult to determine average field widths. From the few available reports on traditional subirrigated fields or the closely related drained or raised fields, a range of perhaps 3 to 12 m seems reasonable, with more widely spaced drains possible in sandy or gravelly soils.[38]

Field length is much more variable. Soil permeability is not a consideration; instead, length is determined by such factors as space needs and cultivation practices (for example, whether fields are worked by hand or team). Reports of subirrigated, drained-field, and raised-field lengths in the New World range from a few meters to more than 2 km.

Most subirrigated fields are rectangular, although there is no compelling hydraulic reason for this. It may be that construction of interconnecting drains encourages rectilinearity. Raised fields in South America exhibit a variety of linear and curvilinear shapes, all the while adhering to the general rule of narrow configuration.[39]

Construction

Because accounts of new subirrigated-field development are rare, a representative example from southern Tlaxcala, Mexico, must suffice. The region lies at the northern end of the great Puebla-Tlaxcala Basin at an elevation of about 2,200 m. High water tables, bogs, and open swamps plus soils with poor internal drainage and underlying layers of impermeable *tepetate* are common throughout the region. Subirrigation is widespread, in combination with furrow and splash irrigation.

New subirrigated plots are created by extending drainage canals into the difficult environments of boggy soils *(suelos pantanosos)* and open swamps *(ciénagas)*. Around individual plots straight-sided *zanjas* a meter or so wide and 1 to 2 m deep are customary. (In the example here, *zanjas* are one meter wide and deep.) Drain excavation is entirely by hand; farmers use only flat-bladed spades or occasionally crowbars to dislodge rocks or break up *tepetate*.[40]

[38]William M. Denevan and B. L. Turner II, "Forms, Functions, and Associations of Raised Fields in the Old World Tropics"; James J. Parsons and William M. Denevan, "Pre-Columbian Ridged Fields"; B. L. Turner II, "Prehistoric Intensive Agriculture in the Mayan Lowlands"; Gene C. Wilken, "Drained-Field Agriculture."

[39]William M. Denevan, "Aboriginal Drained-Field Cultivation in the Americas."

[40]On recent reclamation projects the government used draglines to open new *zanjas madres*. Individual field drainage, however, remains the owner's responsibility and is done by hand.

Figure 10-5. First excavation of a *zanja* to depth of about 40 cm. Note *tule* stalks and other bog vegetation.

Excavation of *zanjas* proceeds in three steps along previously marked plot boundaries. A first cut with the spade removes material to a depth of about 40 cm. This cut takes out most of the vegetation, root systems, and other surface debris. Excavated material is tossed on the plot, where it will partially dry and later be broken up, spread, and eventually plowed (Figure 10-5). A second cut, also of about 40 cm, lowers the trench slightly below the water table, which in our example was at 70 cm (Figure 10-6). Excavation is now more difficult; the worker must stand in water and lift the heavy, saturated material clear of the deepening *zanja* wall. This material, black and permeated with partially rotted bog vegetation, is also added to the new field surface. In the third and final step the last 20 cm of muck is scooped, leaving about 30 cm of standing water in the drain. A final dressing of the sides and bottom with the spade completes the task (Figures 10-7 and 10-8).

In this example, the worker was also the owner *(dueño)*, a young man at the height of his strength, breaking out his first piece of land. The combination of youthful vigor and strong motivation resulted in an excellent performance. The first two steps proceeded at the rate of about 2 m (linear)/hr, or about 1.6 m^3 ($2 \times 1 \times 0.8$)/hr. The third step, the most difficult, also proceeded at the rate of 2 m/hr, or 0.4 m^3/m. Thus, total rate of advance is about one linear meter, or one cubic

Figure 10-6. Second excavation of a *zanja* to depth of about 80 cm. Bottom of drain is now below water level.

meter, per hour, a figure that compares favorably with other manual excavation work performed elsewhere in Mexico. Farmers recognize that hired workers are less motivated and perhaps less capable. Work rates *(tareas)* in this type of mucky, root-filled soil are about 5 m^3/day. If hard *tepetate* is encountered rates can drop to 3 m^3/day.

Although there is considerable variation in plot size, typical dimensions might be 10 m wide and 50 to 100 m long. Choosing a midpoint of 75 m would give a perimeter distance of 170 m. Because *zanjas* customarily line the two long sides and at least one of the short sides, initial drain length is 160 m. At a conservative 4 to 5 m/day, excavation would require 30 to 40 man-days, a not-exorbitant investment for a productive, subirrigated field.

Usually perimeter drains are sufficient to lower plot water tables to acceptable levels. Occasionally a particularly wet field, or one underlain

Figure 10-7. Completed *zanja*. Excavated material on left will be spread over new plot.

Figure 10-8. New drained field planted to maize, husk tomatoes, squash, and other crops. Unreclaimed *tule* bog in background.

by a seep or spring, requires additional *zanjas*. These are simply led into the field from the closest edge directly to the wet spot. They constitute a nuisance not so much because they require additional effort to excavate as because they disrupt plowing and other field activities. In some cases I have seen such midfield drains covered with rock and soil so that teams could pass directly over them.

Some benefits, such as partial drainage, are derived as soon as one of the *zanjas* is completed. But in most boggy soils the full complement of *zanjas* is needed to adequately drain the field in wet periods and supply subirrigation water in the dry season. Work may proceed slowly, however, as time to excavate or money to pay hired labor are available—a rate of progress that suits the small-scale farmer. Furthermore, new fields are of unknown qualities. Exact drain spacing is uncertain,[41] and soil characteristics, and thus internal drainage rates, are unknown. Farmers prefer to proceed cautiously, perhaps partially draining a field, then trying a few crops experimentally the first year to get the "feel" of the new plot before committing additional resources to reclamation.

Installation of drainage works does not end field preparation. Excavated material must be distributed and plot surfaces leveled. In the example total volume of excavated material was 160 m^3, which when spread over the 10-×-75-m plot would elevate the surface only about 20 cm. Although additional material—for example, river sand, which also improves the tilth of heavy bog soils—is occasionally brought in, plot heights are usually not further adjusted. Instead, if water levels prove to be too high, drains are deepened. The goal is a planting bed far enough above the water table so that plants are not jeopardized by saturated soils. Only deep-rooted crops such as mature maize or alfalfa tap the capillary fringe; shallow-rooted vegetables and young field crops are surface irrigated. Soil fertility is maintained, in part, by additions of muck *(lama)* from the *zanjas*. These amendments contribute little to plot height because they are derived largely from material eroded from the plots themselves.

Thus, traditional Middle American farmers do not achieve true subirrigation in the sense of managing the relationship between plot surface and water tables. Instead, subirrigation is almost a by-product of reclaiming wet or boggy fields. Direct surface irrigation substitutes for the careful adjustment of water tables in response to changing crop conditions.

Farmers who would lower planting surfaces to take advantage of subsurface water must also face the problem of coordinating capillary

[41] D. B. Trafford ("Agricultural Land Drainage") notes that in Europe drain spacing was a subject of debate "capable of arousing fierce passions" and that the many rules of thumb were "eloquently argued by their supporters and soundly denounced by their detractors."

fringes and root zones. There are few examples of this technique in Middle America.[42] For example, a system of excavating small pits 20 to 30 cm deep and planting maize in the exposed moist soil reported for the Teotihuacán Valley more closely resembles the dune farming of the Hopi than subirrigation.[43] Even this limited practice is almost prohibitively laborious as evidenced by its name, *a todo costo* (at all costs or effort).

Extent

Muddling the distinction between traditional subirrigation and drainage makes it difficult to define extent and distribution of the various practices. Perhaps a few examples will suggest their widespread and varied application. Father Alonso Ponce implied a subirrigated basis for the *chinampas* as early as the sixteenth century:

> As these gardens are raised less than a vara [84 cm] above the water, even without rainfall they bear vigorous maize, sustained by the moisture provided by the lagoon.[44]

Most modern writers agree with this description, although noting that surface irrigation is necessary for seed beds and young plants and in dry spells.[45]

In the central valleys of Oaxaca Kirkby found a transition in type of irrigation based on ground-water depth: from water-table (subirrigation) farming at 0.25 to 1.0 m, to marginal ground-water farming at 1.0 to 2.5 m, to pot irrigation *(riego a cántaro)* up to 8 m deep.[46] Subirrigated plots are hand watered if newly planted or if the water table falls.

In highland Guatemala near Antigua *zanja*-bordered plots strikingly similar to the subirrigated fields of Tlaxcala probably derive some of their moisture from the water table. Farther west, fields on the shores

[42]The sunken fields *(mahamaes, pukios)* of coastal Peru are generally thought to be subirrigated. Jeffrey R. Parsons, "The Archaeological Significance of Mahamaes Cultivation on the Coast of Peru"; Jeffrey R. Parsons and Norbert P. Psuty, "Sunken Fields and Prehistoric Subsistence on the Peruvian Coast"; Michael West, "Early Watertable Farming on the North Coast of Peru."

[43]William T. Sanders, "Tierra y Aqua (Soil and Water)," p. 125; John T. Hack, *The Changing Physical Environment of the Hopi Indians of Arizona,* pp. 32–34.

[44]Translation in Pedro Armillas ("Gardens on Swamps") of [Alonso de San Juan], *Relación Breve y Verdadera de Algunas Cosas de las Muchas que Sucedieron al Padre Alonso Ponce en las Provincias de la Nueva España.*

[45]Michael D. Coe, "The Chinampas of Mexico"; Robert C. West and Pedro Armillas, "Las Chinampas de México."

[46]Kirkby, *The Use of Land and Water Resources,* pp. 48, 50–51.

of Lake Atitlán may be low enough so that maize and other deep-rooted crops can tap the capillary fringe.[47] At Almolonga and Quezaltenango plots reclaimed from swamps stand only 40 to 70 cm above the water level in surrounding canals. But the primary crops here are shallow-rooted vegetables, and the *zanjas* serve mainly as drains and sources of water for splash irrigation rather than for subirrigation.

The situation in the eastern lowlands of Mexico and perhaps elsewhere in Middle America remains unresolved. Extensive ancient raised-field systems are coming to light with startling rapidity.[48] Most of these plots exhibit the long narrow shape associated with subirrigation and drainage, which is not surprising in a region where rainfall is high and lands are often inundated. But the fields have been abandoned past recollection, and modern use is limited mostly to grazing. Considering the variety of uses to which raised fields are put, it is perhaps premature to speculate on the original methods of operation and irrigation.

These few examples only suggest the almost pan-regional range of subirrigation and drainage in Middle America. As is true in the rest of Latin America, raised, drained, and subirrigated fields were one of the major agricultural engineering achievements of the past.[49] The techniques continue and are even being promoted.[50] Costs may be a deterrent; labor requirements for construction (investment) and maintenance (operating costs) are formidable. For example, at the excavation rate of 4 m^3/day developed earlier, it would require something on the order of a man-year of work days to install meter-wide and meter-deep drains on a hectare of land divided into ten plots each 10 × 100 m. Maintenance, which is absolutely necessary to keep drains open and free of choking silts and aquatic vegetation, demands additional labor.

But apparently traditional farmers have been willing to make these investments and pay these costs to bring otherwise unusable land into production. Operating and relic drained and subirrigated fields cover tens of thousands of hectares in Middle America. The environment encourages their development; high water tables and poorly drained soils abound in both the highlands and the lowlands. And perhaps the dense populations demand them; most areas of past and present drained

[47]Gene C. Wilken, "Food-Producing Systems Available to the Ancient Maya."

[48]For reviews of these studies, see William M. Denevan, "Latin America"; Norman Hammond, "The Exploration of the Maya World"; Peter D. Harrison and B. L. Turner II, eds., *Pre-Hispanic Maya Agriculture.*

[49]Denevan, "Latin America"; Denevan, "Aboriginal Drained-Field Cultivation in the Americas."

[50]Alfredo Becerril Colín, "Proyecto Preliminar de la Planificación de la Cuenca del Valle de México"; Arturo Gómez-Pompa, "An Old Answer to the Future."

fields are associated with dense populations that consume the products and supply the labor to sustain intensive drained-field agriculture.

Silts and Salts

Silts and salts have posed threats to traditional water managers in all places and times.[51] Silts are a mixed curse. They are essential to silt-trap terrace formation and serve as valuable field dressings in *zanja/pantle* and drained-field systems. But they also clog canals and drains and must be removed at heavy cost. In extreme cases when cleaning silt deposits from canals, reservoirs, and aggrading stream channels requires mass transfers that exceed the resources of traditional communities,[52] entire water-management systems are abandoned.

Salts, however, present a chemical rather than a mechanical problem and are the more difficult for it. Traditional solutions do not differ much from those employed in advanced societies, and they are no more effective. Physical treatments such as subsoil- or deep-plowing require powerful tractors. But traditional farmers do add sand to salty fields to improve permeability and leaching. Alternatively I have seen farmers remove a layer of salt-permeated soil from the surface of their plot.[53] Such solutions, with their high labor costs, are practiced only on small plots in land-scarce regions.

A less energy-expensive solution is to wash or leach the soil with additional quantities of irrigation water. The practice is common wherever salt poses a threat and drainage is good. Real problems arise where drainage is impeded—for example, when drainage networks around subirrigated fields are not maintained. In general traditional Middle American farmers are sensitive to the dangers of salinization and spare no effort to remove salt even under difficult conditions. But when all efforts fail, as they sometimes do, lands are abandoned for cropping. Fields revert to salt-tolerant grasses and become pastures for domestic animals. In essence this is a long-term fallow treatment that over time may reduce salts in soils.

[51]Theodore L. Downing and McGuire Gibson, eds., *Irrigation's Impact on Society;* Eric P. Eckholm, *Losing Ground;* Thorkild Jacobsen and R. McC. Adams, "Salt and Silt in Ancient Mesopotamian Agriculture."

[52]They exceed the resources of industrial societies, too. Even with powered equipment it is often uneconomical to dredge deposits from reservoirs and flood-control catchments. The projected life of a dam is an estimate not so much of when the structure might fail as of when silting will destroy its utility.

[53]Axel Steensberg ("Comments") repeats the Middle Ages Arabic account of thousands of Zinj slaves who "had to remove the surface layer of salts and to carry the waste material away from the irrigated plots. After the Zinj revolted in A.D. 869, 50,000 hectares of irrigated land were abandoned and ultimately became salt marshes," p. 403.

Summary

Several solutions to the problem of extracting subsurface water have been examined here. The deep well pumps and *sociedades de aqua* of Tepeaca are striking examples of a traditional management organization that has been successfully adapted to a new technology. The *galerías filtrantes* are a quintessential traditional solution: they require little in the way of special tools or equipment to construct; they can be built intermittently, as funds and labor are available; they can be operated and maintained locally; and although costly to build, they deliver water at much lower labor cost than manual systems.

Similarly, the subirrigation and drained-field systems provide excellent examples of the traditional approach. Field surfaces are modified so that natural processes, in this case capillary action and drainage, help rather than hinder the farmer's efforts. Since subsurface water levels cannot be controlled, farmers do not achieve true subirrigation. Nevertheless, the *chinampas* and drained fields reviewed here are effective responses to the challenge of high water tables.

11

Climate Management

Of all the agricultural resources climate seems the least likely candidate for management. If the enormous energy and moisture fluxes of macroscale or even mesoscale weather phenomena[1] still defy control by even the most technologically advanced societies,[2] what should we expect from traditional farmers? At the same time, no other resource so sets limits for farming or imparts such an element of uncertainty. The middle ranges of temperature and moisture—that is, normal or average conditions—establish basic crop mixes, cultivation practices, and schedules of seasonal and annual activities. Extreme events, such as droughts or hailstorms, can ruin a crop and spell disaster for farm life. Little wonder that management of the climate is so often delegated to the deities. When confronted with the awesome inevitability of climatic events, traditional farmers, no less than their industrialized counterparts, can only adjust planting dates and relieve the anxiety of impotence with ceremony.[3]

[1]T. R. Oke (*Boundary Layer Climates*, p. 3) offers some values for common horizontal distance scales: microscale, 10^{-2} to 10^{3} m; local scale, 10^{2} to 5×10^{4} m; mesoscale, 10^{4} to 2×10^{5} m; macroscale, 10^{5} to 10^{8} m.

[2]National Science Foundation, *Weather Modification, Tenth Annual Report, 1968*, p. 1; F. Prohaska, "Weather Modification in the Tropics for Agricultural Purposes."

[3]Alvin W. Gouldner and Richard A. Peterson, *Notes on Technology and the Moral Order*, pp. 34–35. Anne V. T. Kirkby (*The Use of Land and Water Resources in the Past and Present Valley of Oaxaca, Mexico*, pp. 54–59)

But despite the seeming inevitability of the weather, traditional farmers are not limited to simply reacting to it. Several elements of climate can be managed either to modify (or to extend) normal conditions or to buffer extreme events. The remedy lies in reducing scale, one of traditional agriculture's basic strategies. In the intimate environment of a small plot or single planting bed natural forces are brought to manageable levels, and farmers can implement a variety of climate-control practices that would be impractical in larger spaces. Even at this level farmers rarely apply energy (for example, heat) directly. Instead, they meliorate or redirect natural radiation and moisture using detailed knowledge of environmental processes and local materials.

Before reviewing climate-control measures and their functions, I will briefly examine the microenvironments within which traditional farmers work. Most "climate" is measured in standard instrument shelters positioned 1.5 m above the ground and away from structures and vegetation to avoid local influences. Thus, measurements of temperature and humidity are taken in the zone occupied by the upper part of a standing man. But most crop plants spend their entire lives well below this level, and all pass the critical stages of early growth at or near the ground.[4] This is the active surface, a scene of intense interactions, where solar radiation is absorbed or reflected, heat and moisture flow between soil and atmosphere, and surface friction slows airflows.[5]

Slow transfers of heat and moisture within the soil and the lowest (laminar) layer of the atmosphere as well as reduced wind speeds near the surface permit great differences of temperature and humidity to develop over short vertical and horizontal distances. These steep gradients and the smallness of scale are the distinguishing features of microclimates, which are customarily identified with the surfaces from which they derive their special characteristics. Thus, one speaks of the microclimate of a forest, soil, wall, and so on.[6] A crop zone contains a

---

discusses adjustment of planting dates as a strategy against climate variation. Traditional weather prediction and other environmental strategies are reported from Africa but have not been summarized for Middle America. I. J. Jackson, "Traditional Forecasting of Tropical Rainy Seasons"; Gene C. Wilken, *Agroclimatic Hazard Perception, Prediction, and Risk-Avoidance Strategies in Lesotho*.

[4]Rudolf Geiger (*The Climate near the Ground*, pp. 1–2) distinguishes between the "human climate" at standard shelter heights and the "plant" or "habitat climate" at lower levels.

[5]Oke, *Boundary Layer Climates*, p. 31.

[6]R. M. Holmes and A. Nelson Dingle, "The Relationship Between the Macro- and Microclimate."

number of microclimates, which are sometimes grouped together under a term such as *crop climate* or *ecoclimate.*

Shielding plant spaces is especially important because turbulent winds will stir the air and destroy the steep gradients. Thus, management of microclimates has two aspects: generation of desirable heat and moisture conditions and preservation of these conditions within the crop zone.[7] The mechanical effects of wind, rain, and hail lie outside the processes discussed. But because protection from wind pressure and the impact of falling rain and hail belong in a discussion of microclimate management, a few shielding practices will be considered here.

Any review of climate management poses problems of organization. The climate elements themselves so interact that one invariably changes others. For example, altering radiation levels also affects temperature and evaporation. At times it is almost impossible to determine which change is most important in the crop environment. In some cases multiple changes complement each other; in others they conflict.

Yet to focus on management methods and their effects risks unwelcome repetition for the same reason: almost every practice modifies several elements. A rigorous linking of each practice with all effects would exceed the reader's patience and the understanding most traditional farmers have of the meterological forces at work in their fields. As a compromise, I shall briefly review the major elements but emphasize specific management procedures, relying on the farmer's stated reasons for using particular methods and on my own identification of the most important results.

Major Elements

Radiation

Despite the fact that Middle America lies mostly within the energy-rich tropics, the first problem faced by cultivators is getting enough sunshine. Most crop plants are low-growing heliophytes that require high levels of radiation and temperature to develop and produce. But except in arid areas natural vegetation forms a canopy that intercepts much of the solar radiation well above the crop plant zone. In humid areas dense forest canopies intercept more than 90 percent of the radiation, effectively deactivating the soil surface. Even under less-dense stands radiation levels are substantially reduced.[8] The result is that

[7]Even greenhouses serve more to maintain local climates against turbulent mixing than to alter radiation fluxes. Oke, *Boundary Layer Climates,* pp. 218–20; M. L. Blanc et al., *Protection Against Frost Damage,* p. 31.

[8]Geiger, *The Climate near the Ground,* pp. 270ff.

extensive farming in forested areas is impossible until forest microclimates are replaced by crop microclimates.

The obvious solution is devegetation to eliminate overstories and transfer active surfaces to ground levels. The notorious slashing and burning of shifting cultivation are first of all a means of reducing shade and only secondarily a means of eliminating competitive vegetation and generating nutrients from ashes. Permanent farm lands have also gone through initial deforestation to create the high-energy field surfaces required by most crop plants. However, very high radiation levels and resulting heat loads lower productivity or reduce the quality of some crops and may prove lethal to seedlings. The objectives, therefore, are to achieve adequate but not excessive radiation levels and to adjust them at various stages of plant growth.

In areas where shifting cultivation is practiced a few trees are often left standing in cleared plots. These may be economic or ceremonial trees (for example, *ceiba, Ceiba pentendra*) that are deliberately spared or simply trees that are too difficult to remove. In either case the remnant trees provide partial shade, which protects soils, suppresses weed growth, and improves the crop environment.[9] Following a year or two of cultivation, plots begin their fallow period but are not abandoned. Perennial cultivated or volunteer crops continue to produce and are harvested. Some farmers plant special shrubs and trees in fallowed plots as a continuing source of products and to influence the composition of regenerating stands. These postcultivation crops are usually those that flourish in bush and forest microclimates.

Shifting cultivators are usually regarded as exploiters of favorable environments and beneficiaries of inadvertent changes rather than as deliberate resource managers. For example, in their small forest openings wind speeds are low and thus evaporation is reduced; in addition, precipitation is increased because rain falls out of the airstream in the eddies over their plots.[10] But the farmers also take an active role in managing their microclimates. By clearing the vegetation they create high-energy field surfaces for their crops. By leaving a few trees they provide partial shading and encourage regeneration of forest environments.

Trees are used to reduce radiation loads in permanent field systems too. They may be remnants of original forest or, more likely, planted specifically for shade. Best known in Middle America are the "nurse" *(criada)* or "mother" *(madre)* trees used to shade the beverage crops, cacao and coffee, and the shrubs and low trees that support and shield vanilla and pepper. These crops appear to be natural understory plants,

[9]Emil B. Haney, *The Nature of Shifting Cultivation in Latin America.* Harold C. Conklin, *Hanunóo Agriculture,* p. 62.

[10]Geiger, *The Climate near the Ground,* pp. 352ff.

Figure 11-1. Multistoried commercial field: shade trees over bananas over coffee, Veracruz, Mexico.

Figure 11-2. Lettuce leaves tied to shield heads from sun.

which generally thrive or produce a higher quality seed in a shielded environment.[11]

Shading with trees is one of the few traditional techniques that has been investigated by modern science, incorporated into commercial plantations,[12] and diffused to other areas. It also illustrates the difficulties of evaluating the management of environmental elements that interact with each other. Shade trees not only lower radiation levels and influence temperature in the understory but also reduce wind speeds and evaporation, intercept precipitation, and compete for soil nutrients and moisture. Some of these effects benefit crop plants but some do not, and it seldom is a simple matter to determine net gains and losses.[13]

Overstory plants may be exclusively for shade or may themselves be crops. In Middle America bananas, *guaje (Leucaena esculenta), guamúchil* (*Pithecellobium* spp.), and papaya are often found over shrub (for example, coffee) and ground crops. With the addition of more and more plants, fields begin to resemble forests, replicating not only the microclimates but also the moisture and nutrient flows of original stands (Figure 11-1).[14]

Some crops can be made to shield themselves. In the commercial vegetable gardens of Puebla and Tlaxcala, Mexico, cauliflower leaves are tied together to keep heads white and thus more marketable. Similarly lettuce is tied to keep leaves light colored and heads tight (Figure 11-2).

On a large scale an overstory of trees is the most practical way to reduce radiation loads on farm fields. But in seed beds *(almácigas, sementeras)* or tiny plots other options are available that permit finer control of sunlight. Light grass *(pasto)* or straw *(paja)* mulches reduce radiation and heat levels on newly planted surfaces and also inhibit moisture losses and absorb the kinetic energy of falling rain and hail (Figure 11-3). Some emergent seedlings, such as chile, coffee, and tobacco, require continuing protection. Over these, farmers construct arbors of branches leaning together or of branches or leaves supported by stick frames (Figure 11-4). Usually an entire seed bed is covered, although in some cases small arbors protect individual seedlings in the field (Figure 11-5). All materials are of local origin. Creating precise

[11]D. B. Murray, "Crop-Weather Relations—Cacao."

[12]G. W. Hurst and R. P. Rumney, *Protection of Plants Against Adverse Weather*, pp. 11–12. Commercial use of shading trees has also resulted in one of the few instances of substantial research and publication on a traditional technology.

[13]Some officials at Mexico's coffee research institute (IMCAFE) argue against certain shade plants, such as bananas, which they feel drain soil nutrients. But the practice continues, not only in traditional gardens but also in commercial plantations.

[14]For other examples of forest replication, see Gene C. Wilken, "Integrating Forest and Small-Scale Farm Systems in Middle America."

Figure 11-3. Light surface mulch, San Agustín Atzompa, Puebla, Mexico.

Figure 11-4. Coffee seedling arbor, Veracruz, Mexico.

Figure 11-5. Shading arbor over individual coffee seedling.

amounts of shade is possible by varying branch or leaf spacing and density. As seedlings grow, they are acclimated to the full sun conditions of open fields by progressive removal of shade materials or by natural withering and falling of arbor leaves.

The *chilares* of Puebla, Mexico, illustrate the procedures and details of one arbor system. After seed beds about 140 cm wide and of variable length are carefully prepared, chile (*Capsicum* spp.) seeds are sprinkled on the surface and covered with a light layer of earth *(tapa con tierra)*. Then branches from almost any convenient source, such as *asomiate (Sencio salignus)*, *jarra* or *jarilla* (here *Baccharis glutinosa)* or *pirú (Schinus molle)*, each about 150 cm in length, are laid flat across the seed beds. The branch-and-dirt mulch protects the surface and seeds from frost, hail, heavy rains, and birds, and reduces soil moisture losses.

After a month the branches are taken up and excess earth is removed *(raspar, sacuda de tierra, destapar)*. Then branches are replaced at 25- to

Figure 11-6. Erecting a chile arbor, San Lucas El Grande, Puebla, Mexico.

50-cm intervals along both sides of the seed beds with their bases in small V-shaped trenches and their tops enmeshed over the seed beds (Figures 11-6 and 11-7). The work goes quickly; a 35-m-long seed bed can be covered (*enjacalada;* from the Nahuatl *xacalli,* literally placed in a hut) in half a day by an experienced worker.

The delicate chile seedlings continue to develop in the benign arbor environment, shaded and shielded from sun, wind, rain, hail, and birds. After two months, during which the arbor deteriorates and plants are progressively exposed to the elements, seedlings are transplanted to open, permanent fields.

In higher latitudes, especially early in the growing season when the sun is low, the problem may be too little rather than too much energy. Some traditional farmers darken the soil to increase absorption of radiation and alter the surface to increase the angle of incidence. For example, east/west aligned furrows offer a sunny south face for early planting.[15] Ridges and walls not only catch low-angle solar rays but also act as windbreaks and extend local microclimate conditions upward into the zone of shrubs, vines, and low trees. These techniques, however, are not common in low-latitude Middle America, where sun angles are

[15]W. C. Burrows, "Characterization of Soil Temperature Distribution from Various Tillage-Induced Microreliefs"; Oke, *Boundary Layer Climates,* pp. 201–2.

Figure 11-7. Chile arbors, San Lucas El Grande, Puebla, Mexico.

high. Similarly long-wave (back) radiation from trees or other nearby objects is of little advantage in the humid tropics, where nights are comparatively warm.[16]

Heat and Moisture

Direct Application of Heat

Although smudging is an ancient and widespread practice, it is not common in tropical small-scale systems. Traditional farmers generally lack the concentrated fuels necessary to alter temperatures in open spaces, especially in the more settled areas, where temperature-sensitive crops are often found. But the practice is not unknown. For example, when frost is expected farmers near Zaachila, Oaxaca, Mexico, sometimes rise at 3:00 in the morning to burn piles of *basura* (literally, garbage; actually field cleanings of leaves, weeds, and crop residue) on the edges of tomato fields. The burning piles generate heat, but perhaps more importantly they generate both smoke, which traps outgoing

[16]Techniques found elsewhere for managing short- and long-wave radiation are reviewed in Gene C. Wilken, "Microclimate Management by Traditional Farmers."

radiation, and turbulence, which mixes stagnant, radiatively cooled air near the ground with warm air aloft.[17]

Field configurations in this area favor effective use of these limited fuels. Plots are small and flat, with no gradients down which cold air could drain away, and they are bordered by trees that serve as windbreaks but also tend to pond cold air on still nights. Thus, the fields are partially enclosed spaces in which the effects of burning are concentrated.

Heat and Moisture Flows

Heat and moisture flows within the soil and between the surface and atmosphere are considered together because they are closely linked and both usually respond to the same management practices. Heat is transferred between surface and substrate by conduction and to some extent by water movement. Thermal conductivity and thus the rate of heat flow are in turn determined by soil porosity, moisture, and organic content.[18] Liquid water in the soil is pulled downward by gravity and in all directions by capillary action. Heat and water vapor move by conduction and diffusion within the first few millimeters of the atmosphere (laminar layer). Thereafter, fast convective and advective exchanges are vastly more important.

Except for direct additions of heat (rare) and water (irrigation) traditional farmers manage heat and moisture levels by controlling flows. On a small-farm plot, surface/atmosphere interfaces are accessible and manageable even by farmers with limited energy resources. The most common approach is to alter the nature of the surface to replace fast, turbulent processes with slower ones. Usually flows in both directions are affected equally. For example, an insulating mulch inhibits conductive flows both into and out of the soil. In other cases controls are one way; the same mulch that reduces upward transfers of water vapor to the atmosphere may not appreciably inhibit downward movement of rainwater.

Control Measures

Tillage

The main purposes of plowing and cultivation are to loosen the soil for root development, to control weeds, and to increase infiltration of

[17]Blanc et al., *Protection Against Frost Damage,* p. 2; G. Z. Ventskevich, *Agrometeorology,* p. 31.

[18]Jen-hu Chang, *Climate and Agriculture,* p. 89; Norman J. Rosenberg, *Microclimate,* p. 65.

precipitation. But loosening the soil also reduces thermal conductivity, and consequently heat flows between surface and substrate. Tillage also breaks capillary connections in the soil and inhibits water transfers to the surface. The tilled layer dries out quickly by evaporation, but subsoil moisture is conserved. Although the efficacy of shallow tilling to conserve moisture (dry- or dust-mulching) has been questioned,[19] farmers continue the practice in many parts of Middle America.

In the sun- and rain-rich humid tropics, tillage that conserves heat and moisture is not of much value, which may partially explain the success of nonplow agriculture in these regions. In fact, some tropical soil management practices are aimed at *lowering* heat and moisture levels. For example, raised planting beds or mounds reduce soil moisture through improved drainage and may also control soil temperatures.[20]

Mulches

Mulches offer a wide range of possibilities for modifying surface characteristics and crop microclimates. Traditional organic mulches are essentially active surfaces supported above the soil by poorly conducting structures consisting of plant remains and entrapped air. These structures partially uncouple the soil/atmosphere interface and affect kinetic, radiative, and turbulent exchanges to a degree dependent on mulch characteristics, depth, and extent of coverage.[21]

Mulches shield soil and seedlings from the physical impact of wind, rain, and hail. Organic mulches block the force of falling raindrops but not the mass; they intercept only a fraction of total precipitation and improve infiltration into the soil. In fact, mulching before the start of rains to improve reception rather than after to suppress evaporation has been recommended.[22] Even the small interception losses are largely offset by high humidity within the mulch, which lowers vapor gradients and further reduces moisture losses from the soil.

Mulch color in part determines reflectivity and thus the amount of radiation available for heating and evaporation. Traditional Middle American mulch materials, such as straw and leaves, absorb and reflect radiation in about the same proportions as do bare soils. But dark

[19]Nyle C. Brady, *The Nature and Properties of Soils,* pp. 178–81; G. V. Jacks, *Soil,* p. 60.

[20]Temperatures on the tops of sizable (3.18 m diameter × 0.6 m high) New Guinea sweet potato mounds are 2°C higher than on cleared, unmounded ground, which allows cultivation at higher elevations than would be possible otherwise. Eric Waddell, "How the Enga Cope with Frost."

[21]J. W. Davies, *Mulching Effects on Plant Climate and Yield.*

[22]Davies, *Mulching Effects,* p. 43.

plastic sheeting is increasingly popular as a cover, especially in highland regions, suggesting that there are advantages to hastening soil heating even in the tropics.

The greatest impact of mulches, however, is in moisture conservation. Because water evaporated from a bare soil surface is replaced from the substrate, losses continue until the whole soil profile dries. But the loose composition of an organic mulch inhibits turbulent transfers and restricts water flows from the soil to the atmosphere to slow diffusion transfers. Evaporation suppression depends on mulch thickness and composition as well as on atmospheric and soil conditions. Although the many variables produce mixed results, Davies reports experiments in which as little as 3 mm of mulch had measurable effects on moisture losses, and 15 cm prevented most evaporation from the soil surface.[23]

Similarly, heat transfers proceed slowly in the poorly conducting materials of mulches and in the still air within them. Temperature gradients between surface and atmosphere are preserved, and covered soil remains cooler than bare soil during daily or seasonal hot periods but warmer during cool periods. The net effect is to lower average soil temperatures while reducing fluctuations and the incidence of frost. On the disadvantageous side, temperature extremes immediately above mulches are increased so that emergent seedlings are at greater risk from both high and low temperatures.

Middle American farmers use a great variety of mulch materials to cover seed beds and newly planted ridges and mounds. Straw from wheat, oats, and barley is popular, as are many kinds of crop residues and grasses gathered from fallow fields and nearby hills. In areas near shallow water, aquatic plants such as water hyacinth *(Eichornia crassipes)* and duckweed (*Lemna* spp.) are pulled from the water and spread over fields, at first nearly covering the surface, then, as they dry and shrivel, exposing more and more of the surface to direct sunlight. Eventually, most plant mulches are incorporated into the soil by cultivation, where they add to organic content. Other covering materials such as newspaper, cloth, and plastic sheets are increasingly popular with traditional Middle American farmers. They are easy to spread and to adjust should weather conditions change.

Shielding Crops and Crop Climates

Many plants are damaged by exposure to strong or consistent winds and most benefit from the benign microclimates discussed previously.[24]

[23]Davies, *Mulching Effects,* pp. 26–27. Plastic sheeting, of micron width, suppresses almost all evaporation. But it also excludes precipitation and must be removed whenever rain seems imminent.

[24]J. Grace, *Plant Response to Wind;* G. J. Rose, *Crop Protection.*

Figure 11-8. Wind throw on edge of maize field, Puebla, Mexico.

Because winds and turbulent mixing destroy local temperature and moisture gradients, it is necessary to slow the wind or create calm eddies to shelter crops and crop climates. Any of the techniques already reviewed that involve structures of some kind also disrupt airflows. For example, overstories of shading trees and shrubs slow wind movement over fields. Mulches create a layer of calm or dead air above the soil surface in which slow transfers and steep temperature and vapor gradients prevail. But in addition Middle American farmers employ a wide range of trees, shrubs, mats, and walls specifically to shield their plots from prevailing or storm winds.

Height and density are the most important characteristics of windbreaks, assuming that lengths are adequate (on the order of ten or more times height). Significant slowing (20 percent or more) of airflows extends downstream a distance of ten to twenty times windbreak height depending on its permeability. Although a solid or impermeable barrier most nearly blocks winds, it creates great turbulence. Semipermeable barriers not only slow wind speeds almost as much and generate fewer eddies, but their effects extend farther downstream.[25]

[25]J. M. Caborn, *Shelterbelts and Microclimate;* Geiger, *The Climate near the Ground,* pp. 36–40; A. R. Kostaninov and L. R. Struzer, *Shelterbelts and Crop Yields;* J. van Eimern et al., *Windbreaks and Shelterbelts.*

As was the case with shading, scale is important. Seed beds, small plots, or dooryard gardens may be shielded with brush, woven mats, fences, walls, or hedges. Of the great number of hedging plants used in Middle America *nopal* cactus (*Opuntia* spp.) and *maguey* (*Agave* spp.) are especially popular. They stand up to the fiercest winds, resist animal encroachment, and, although both have edible parts and other useful products, are not particularly susceptible to pilferage.

Large fields need trees or shrubs for windbreaks *(cortinas, rompavientos).* These may consist of otherwise noneconomic species, such as the lines of pine that shelter orange and banana groves from the damaging *norte* and *trifulaca* winds of coastal Veracruz, or may themselves produce edible seeds or fruit.

Alternatively, the crops can shield themselves. Normally the interior of a maize field is well protected from the wind.[26] But lodging (wind throw) on the edges not only allows winds to penetrate the stand but also may create a funnel effect, with accelerated winds flowing through the gaps and causing great damage (Figure 11-8). Maize plots that are not completely mounded may at least have the outer three or four rows reinforced to establish a windbreak for the rest of the field.

Summary

Local climate plays a dominant and much-respected role in the lives and fortunes of farmers everywhere. And, contrary to familiar maxims, doing something about the weather is an almost universal characteristic of farming. But are traditional climate-management practices effective? Certainly they modify critical elements of crop climates, but do the results justify the labor expended?

The question can be answered only partially at this time. Those procedures that buffer extreme events, such as frost or hail, may make the difference between success and total loss, and clearly are worth considerable effort. Those that modify only average conditions of radiation, heat, or moisture are more difficult to evaluate.

Data that would resolve these questions are rare. Except for some midlatitude measurements of the effects of mulches and windbreaks that might be transferable to the tropics, few figures could be applied to traditional plots.[27] Without these, it is difficult to even guess at the

[26]Rosenberg, *Microclimate,* pp. 101ff.

[27]For a notable exception, see the work of C. J. Stigter and associates at the University of Dar es Salaam, Tanzania; they have measured the effects of several traditional management practices and have also explored innovative data-collection methods. *Microclimate Management and Manipulation of Traditional Farming* and "Shading."

effects on crop development and yield. Instead, we are forced to the familiar but uncomfortable position so often taken with respect to traditional technology: if the farmers spend the effort, they must be receiving commensurate returns. This position is condescending because it implies infallible "peasant wisdom" while reserving for industrialized farmers the need to verify results with scientific procedures.

The situation is remediable. Induced changes in radiation, temperature, and moisture are measured routinely in midlatitude agricultural experiments as are effects on crop yields. Because traditional climate management is usually associated with crops grown for sale, results in harvest quantity and quality could be compared with production costs. Such benefit/cost analyses are common in industrialized societies and are sorely needed in traditional small-farm sectors.

12

Space Management

Discussions of space in agriculture commonly deal with the costs of overcoming distances between places. Analyses concentrate on the distances between farms and markets or between fields and farmsteads with the aim of determining least-cost arrangements of these various elements.[1] This negative view of space as obstacle applies when the focus is on the cost of transferring materials and crops. But when we analyze agricultural production space can be considered a resource. Crop plants require certain amounts of horizontal and vertical space above and below ground to extract nutrients and moisture, intercept sunlight, and develop physically. Like other resources, space can be managed to increase production. Although the relationships between farm space and productivity have not been ignored, most research has been limited at the technical level to horizontal plant spacing and at the aggregate level to optimum farm size and economies of scale.

The concept of space as a resource must be placed within the context of traditional farming. Typically holdings are less than 2 ha (5 acres), often divided into two or three parcels.[2] From this slender base come products for subsistence, and, through sale or exchange, other

[1]Michael Chisholm. *Rural Settlement and Land Use.*

[2]Of all holdings enumerated in the last published world census of agriculture, more than 80 percent were smaller than 5 ha. The average small farm contained fewer than 2 ha. Food and Agriculture Organization, *1970 World Census of Agriculture.*

necessities or luxuries the farmers and their families have. Operating on such a small and tenuous scale, traditional farmers are alert to opportunities for using their few resources as completely and as often as possible.

Like other resources, space becomes scarce when demand exceeds supply. In traditional societies scarce space is often the result of increasing population and of land-tenure patterns that allot only a fraction of the arable land to the traditional farm sector. In addition to subsistence crops, commercial-crop requirements or social production may increase the need for farm space.[3]

One well-known response to space shortages is expansion or migration to other lands. Spontaneous and planned settlements in new areas have occurred throughout the history of Middle America. But there are powerful deterrents to this sort of movement, such as the remoteness and marginality of new lands, the security of tenure, and the social dislocation that results.

Alternatively, farmers may devise or adopt methods for using available space intensively; increasing the horizontal or vertical densities of crops or increasing the frequency of field use.[4] The process is interactive because intensifying the use of space places a drain on other resources; of necessity farmers who carefully manage farm space must also exercise considerable control over other elements in crop environments. In addition, as for other resources, management of space takes time and energy; it is undertaken only if apparent benefits outweigh the additional costs. The practices reviewed here are implemented in response to real or perceived space shortages and are not usually found where space is abundant.[5]

An individual farmer's productive space may be limited absolutely or relative to another resource. Thus, a farmer might apply intensive, space-conserving measures on a small irrigated plot but farm extensively on a large, rain-watered field. Similarly, wives may tend dooryard gardens intensively while husbands extensively crop remote fields. Because this arrangement is partly a result of the convenient location of dooryard gardens, the argument would return us to the sort of distance-cost spatial analysis mentioned earlier.

[3]Harold C. Brookfield, "Intensification and Disintensification in Pacific Agriculture."

[4]Ester Boserup, *The Conditions of Agricultural Growth.* B. L. Turner II, Robert Q. Hanham, and Anthony V. Portararo, "Population Pressure and Agricultural Intensity." But Brookfield ("Intensification and Disintensification") rejects any simple relationship between population and intensity of cultivation.

[5]But David A. Norman ("Rationalizing Mixed Cropping Under Indigenous Conditions") finds that mixed cropping also yields a greater return per unit of labor, which suggests that the technique can be economic regardless of the amount of available land, provided cultivation practices remain constant.

In the discussions in previous chapters the universality of the constraint of scale suggests that resource management in general could have as easily been considered the management of space. That is, most of the techniques examined so far could also be interpreted as efforts to increase the outputs of fixed-size farms (or yields per unit of area). But in addition to the practices specific to common resources others deal directly with space. It seems worth risking a questionable category to review them separately. Space-management methods are considered here under two subdivisions: spatial arrangements and scheduling.[6]

Spatial Arrangements

Field Layout and Equipment

Efficient utilization of space begins with general field layout and configuration. The first objective is to minimize nonplanted spaces. But at the same time a certain amount of what might be called field infrastructure is necessary. Farmers must allow room in their plots for themselves, for their power and transportation equipment, and for other activities such as irrigation and threshing.

With the available technology alternatives may be viewed as trade-offs between space and labor or other resources. For example, as a general rule labor-saving equipment requires a lot of space. Mechanized field equipment, such as tractors or combines, do well in open, uniform fields but poorly where space is limited. In contrast, the maneuverable tools and power sources of traditional farming work well even in small, awkward field edges and corners. The ultimate in this regard is hoe cultivation, in which there is hardly a piece of ground too small or irregular to escape cultivation (Figure 12-1).

Similarly, the several ways of transporting materials and harvests between fields and farmsteads vary in their space needs. Using trucks takes minimum human effort but does require space-consuming farm roads. In addition, the products of some of the field space must go to pay for fuel and maintenance.[7] Draft animals do not need much space in the form of roads and trails, and loaded horses, mules, or burros even less. But loading and hauling with them requires considerable human

[6]Kent V. Flannery ("Archaeological Systems Theory and Early Mesoamerica") uses the term *scheduling* to identify a strategy for resolving conflicts over time and labor in hunting/gathering societies. The term here identifies a strategy for resolving conflicts over the use of land in farming.

[7]The argument for strict accounting of energy costs in farming can also be made for space costs. Gene C. Wilken, "The Fallacy of Labour/Land Comparisons of Agricultural Productivity."

Figure 12-1. Precipitous, rocky field, cultivatable only by hoe (planted to maize and beans).

time and effort. In addition, field space is needed for fodder production, or if crop residues are used to feed animals, soils are deprived of organic materials. Human transport uses the least space because footpaths are narrow, often no more than the normal rows between plants. Labor costs, however, are high.

In the fields of a few space-scarce farming regions in Middle America, only narrow footpaths subtract from crop spaces. All field work and transfers of materials are by humans (Figure 12-2). Even irrigation water is delivered to plants by hand, thereby minimizing the need for space-consuming irrigation works. But most traditional systems strike a balance between scarce space and arduous labor by using combinations of humans, animals, and powered equipment. Thus, loads may be carried from field centers to edges by hand, then to farmsteads by animal-drawn carts, and finally to market in trucks.

Labor-saving equipment requires space not only for access to fields but for storage as well. For convenience this storage space is usually set aside near farm dwellings. Farmyards in industrial societies have become machinery depots, large open areas with packed or even paved surfaces used for storing and maneuvering powered or pulled equipment. These most accessible lands, which theoretically should be the most intensively cultivated, are as devoid of life as parking lots, which

Figure 12-2. Farmers carrying harvests from fields, Quezaltenango, Guatemala.

Figure 12-3. Harvested *tule* on drained field.

in a sense they are. In nonindustrial societies house yards are often the most productive parts of farms. When space does not have to be used for storage of bulky equipment, house yards can be devoted to the cultivation of choice, delicate, or experimental plants as well as basic crops.

Alternatively, field structures themselves may be productive. In traditional farming regions field edges and borders, roadsides, and canal banks are routinely grazed by farm animals, although this practice requires constant supervision by herdsmen. Where space is even scarcer, these surfaces may be planted to specialty crops, herbs, or medicinals, or to deep-rooted shrubs and trees that yield valuable products and also act as soil stabilizers and windbreaks. Even waterways can be productive. Where water flows and temperatures are right, the salad green *berro* (watercress, *Nasturtium officianale*) and *tule* (*Scirpus* spp.), used for mats, are planted in canals and field drains (Figure 12-3). Distinctive field-edge and aquatic microenvironments also shelter wild plants and animals that are harvested.[8] Thus, to the techniques for intensively cropping field surfaces must be added possibilities for exploiting other spaces within the farm system.

Spacing

Horizontal Space

Planting intervals in space-intensive systems are not especially close; they correspond to recommendations found in common guides to tropical horticulture.[9] For example, anyone familiar with home or commercial gardening will recognize the spacings in Table 12-1, which are typical of traditional plots in Middle America. The big difference is in row spacing, which in less intensive or more mechanized systems is set by the needs of cultivation tools and is usually two or three times as great as plant spacing. Because hand-tended plots do not need wide rows, row spacing can be the same as plant spacing with resulting high plant densities.

Considering the importance of maize in the diet and economy of Middle America, there are surprisingly few data on traditional spatial arrangements for this crop. Most accounts suggest only that plants and rows are about one pace apart, probably on the order of a *vara* (84 cm)

[8]Gene C. Wilken, "The Ecology of Gathering in a Mexican Farming Region."

[9]Luis Alsina Grau, *Calendario del Horticultor;* Ernest Mortensen and Ervin T. Bullard, *Handbook of Tropical and Sub-Tropical Horticulture;* H. D. Tindall, *Commercial Vegetable Growing.*

Table 12-1. Plant Spacing

Crop	Spacing (cm)[a]
Carrots,[b] onions	10
Celery, garlic	10–15
Spinach	15–20
Beets, leeks	20
Lettuce	25–30
Cabbage, cauliflower	40–50

[a]Most spacings are fractions or multiples of a handspan (*cuarta*), which is about 20 cm.
[b]Spacing varies with size desired, which in turn reflects different market preferences.

to a meter.[10] Although the number of seeds planted in each *mata* is often faithfully reported, information on plant survival and maturation is scanty. Thus the literature ignores spacing and number of mature plants, the two factors that determine plant density. In addition, the data supply few correlations of other important variables such as soil and climate types, irrigation, technological levels and tools, field size, and cropping practices that might affect farmers' decisions about intervals and densities. Such correlations would be useful. In addition to providing traditional appraisals of optimum density and yield with given resource variables, information on spatial arrangements might also offer insights into perceptions of maize fields as a whole. These perspectives will be briefly explored later.

Lacking extensive data, I can make only a few comments based on my own limited observations and measurements. I measured hill and row spacings in fifty-six different maize fields *(milpas)* in central and southern Mexico and western Guatemala. Although hill spacing ranged from 45 cm in an irrigated *milpa* in Mexico to 150 cm in a hand-worked, *temporal* (rain-fed) field in Guatemala, well over half the samples fell between 80 and 110 cm. Row spacing ranged from 55 to 130 cm, and almost 80 percent of the samples fell between 80 and 130 cm. For all fields average hill spacing was 97 cm and row spacing 90 cm, for a density of 11,000 to 12,000 hills/ha.[11]

[10]Team-worked and most hand-tilled fields have linear rows or planting beds. I have seen only a few small plots and dooryard gardens where there were no discernible rows.

[11]A calculated value; many of the measured fields were smaller than one hectare.

I could determine plants per hill accurately for fewer than half the fields. Farmers reported planting from three to seven seeds, and I counted from two to five mature plants or stalk stumps. Calculated field densities range from 20,000 to 65,000 plants/ha, with most fields clustered within a range of 30,000 to 50,000.[12]

High-density fields were irrigated and were located in regions where stalks and leaves are valued for animal forage. In these fields plant populations approach or even exceed those recommended by recent development projects.[13] Although the widest hill and row spacings were encountered in the department of Totonicapán, Guatemala, farmers there tend to plant six or seven seeds per hill and thus achieve plant densities comparable to those in regions with closer spacing.

What are traditional farmers' views on *milpa* spacing? How do they determine the proper spacing for their maize plants? Perhaps the most common reason farmers give for their particular spacing is that maize roots need adequate room to develop and that anything less than traditional spacing will result in inferior yields. This view seems especially strong with respect to *temporal* fields; maize spacings in irrigated and fertilized fields are often considerably less than average. From this perspective maize fields consist of a limited number of cubical soil spaces or root zones, each of which is the minimum size necessary to supply a particular number of plants with nutrients and moisture. Plant spacing and density, therefore, are based essentially on evaluations of soil and climate characteristics.

In addition to these ecological concerns, other cultivation factors affect spacing decisions. Field preparations give structure to planting surfaces and limit spacing options. Distances between furrows and rows formed by team-drawn plows are established by the span between animals, which varies in turn with type of draft animal and type of harness or yoke. Even preparations of hand-cultivated fields establish a surface geometry that sets row spacing. For example, maize is planted down each of the wide, hoe-formed *camellones* (raised beds) of western Guatemala, and thus bed width, determined in part by hoe length, sets row spacing. And, whether team- or hand-cultivated, fields are harvested by hand and must therefore contain room for harvesters. From this perspective maize fields are work places occupied by the plants and the human and other power sources and equipment. Spacing is thus determined largely by task and traffic requirements.[14]

[12]These densities correspond closely to optimum densities determined in test plots. Felix Ramírez P. and R. J. Laird, *Densidad Optima de Plantas de Maíz para los Valles de México y Toluca.*

[13]Centro Internacional de Mejoramiento de Maíz y Trigo, *The Puebla Project 1967-69,* pp. 38-39.

[14]The open subsistence maize plots of Middle America contrast sharply with the dense fodder fields of the United States, where the problem of field access is

Farmers also often mention wind in relation to spacing, arguing that anything closer than the customary distance between hills produces solid rows subject to wind throw (lodging) and that any greater separation is unnecessary. From this point of view maize fields are cubical spaces above the surface that must be permeable or semipermeable in the horizontal plane. The view has merit. Tests of windbreaks elsewhere show that permeable rows of vegetation more effectively reduce wind speed and are less subject to wind pressure than are impermeable rows.[15] At just what spacing mature maize produces such a semipermeable screen, however, is uncertain. One-meter spacing, especially if the maize is mixed with climbing or shrub beans, may be optimal. Farmers also argue that closer spacing would not provide adequate soil surface for hilling around each set of plants. Because hilling protects against lodging, this is simply another expression of spacing in relation to wind hazard.

Finally, microclimates created by the maize itself appear influential in farmers' decisions about spacing. In Middle American latitudes, maize in hills about one meter apart casts only open shade, which does not inhibit the growth and development of intermingled beans, squash, and other crops commonly grown in *milpas*. From this perspective, maize fields are again viewed as cubical spaces that must be permeable or semipermeable but in this instance in the vertical plane to allow solar radiation to reach the ground.[16] Perhaps more important, maize spacing is not a separate consideration but is instead linked to traditional practices of intercropping, the topic to which we now turn.

Intercropping

Growing two or more crops in the same field is common wherever farmers need to maximize total output per unit of area. The many possibilities and combinations have attracted but so far escaped sys-

---

solved differently. Cultivation equipment is raised to straddle young maize plants, and harvesting equipment opens its own way by clear-cutting the entire crop: ears, leaves, stalks, and all. Thus, although powered equipment is bulky, it does not require wide plant spacing because it never enters the plant zone.

[15]J. M. Caborn, *Shelterbelts and Microclimates*; J. van Eimern et al., *Windbreaks and Shelterbelts*; J. K. Radke and R. T. Hagstrom, "Wind Turbulence in a Soybean Field Sheltered by Four Types of Wind Barriers."

[16]Widely spaced maize plants apparently mature faster than those densely planted. Ramírez P. and Laird, *Densidad Optima de Plantas*. R. S. Loomis ("Traditional Agriculture in America," p. 451) adds yet another dimension. In pastured and rotated fields "the value of land lies not so much in its receipt of sunlight or water, which seldom limit production, but in the rate at which it gathers the scarce resource, nitrogen."

tematic classification.[17] In Middle America farmers may plant two or more annual crops simultaneously in alternate rows or intermingled in the same row; annual crops with or under biennial or perennial crops, which may be the same or mixed; short-term annual crops with long-term annual, biennial, or perennial crops during early stages of plant development; or randomly, with no set chronological or spatial arrangement (for example, in dooryard gardens). In practice various combinations are found, and distinctions between techniques and even between spatial and sequential relationships become unclear.

When soil and water are also under careful management, intercropping becomes complex. Instead of two or three crops, a single plot may contain as many as five or six, growing in rows at uniform intervals and frequencies or randomly. Such intercropping is not limited to subsistence plots. Commercial beets, cabbage, cauliflower, carrots, celery, coriander, lettuce, potatoes, radishes, and spinach are grown in combinations designed to maximize production per unit of horizontal space and to ensure against loss should prices on any one be low at harvest time. Food crops share field space with annual and perennial flowers, spices, and herbs, also destined for market.

Vertical Space

Even though vegetables tend to be of similar height, there is some vertical differentiation. Thus, root and tuber crops such as beets and potatoes may share horizontal space but not vertical space with stem or leaf crops such as celery or spinach. Even greater stratification occurs in the subsistence combinations of maize, beans, squash, and other traditional native plants commonly found in one field. Still further exploitation of different vertical planes is possible by combining crops of widely different growth habits, such as economic shrubs or trees with field or garden crops. Thus, orchards of apples, apricots, *capulín (Prunus capuli),* peaches, pears, and *tejocote (Crataegus mexicana)* found on high plateaus and upper slopes of mountains commonly have understories of maize and beans.

Such high and low tree and shrub crops are even more common in the temperate *(templada)* and hot *(caliente)* zones. Shading the beverage

[17]American Society of Agronomy, *Multiple Cropping,* pp. 2–3; Dana G. Dalrymple, *Survey of Multiple Cropping in Less Developed Nations; ICRISAT, Proceedings of the International Workshop on Intercropping;* C. L. Keswani and B. J. Ndunguru, *Proceedings of the Second Symposium on Intercropping in Semi-Arid Areas;* Norman, "Rationalizing Mixed Cropping"; Hans Ruthenberg, *Farming Systems in the Tropics;* M.J.T. Norman, *Annual Cropping Systems in the Tropics,* pp. 137–38. Although the term *intercalate* (Spanish *intercalar*), to insert between others, is no more precise than *intercrop,* it enjoys wide use in rural Middle America.

plants, cacao and coffee, appears to be an ancient practice. Multiple stories of coconuts over papaya over cacao, bananas, or oranges are found all along the central and southern Caribbean coastal plain of Mexico. Field and garden crops such as maize, *yuca* or *casava (Manihot esculenta),* and beans are planted under these tree-and-shrub combinations to create a four- or even five-layer crop mix in which none of the separate canopies compete for space. However, because there is competition below ground, these multiple combinations are usually limited to those areas where neither precipitation nor solar energy is lacking.

In some intercropping systems combinations of high and low plants are modified to accommodate different growth stages. Thus, maize and other field crops are commonly planted between rows of young mango *(Mangifera indica)* and avocado *(Persea americana).* Later, expanding canopies restrict ground crops to those that tolerate shade better than maize does.

Spacing and sequencing combinations are many; possible crop combinations are endless. The mix of maize and beans (Nahuatl *tlatlaoyo*), often accompanied by squash, has come to typify traditional garden fields throughout Middle America. But that familiar trilogy is only one of the common field mixes. Other crops, such as chayote *(Sechium edule),* red and husk tomatoes (*Lycopersicum esculentum* and *Physalis* spp.), *quelite* (*Chenopodium* spp.), broad beans *(Vicia faba),* and *ayocote (Phaseolis coccineus),* are often found in *temporal* fields along with the maize/bean staples. Volunteer crops from the previous season complicate deliberately planted mixes but are by no means precluded from the harvest.

Dooryard Gardens

The fullest exploitation of vertical as well as horizontal space is found in the multistoried dooryard gardens of Middle America. Plots of no more than 0.1 ha may contain two dozen or more different economic plants, each with distinctive space requirements.[18] A garden in the *templada* region near Atlixco, Puebla, is an example. The plot displayed a seemingly random horizontal distribution but a rather carefully arranged four-tier vertical arrangement of tall trees—mango, papaya *(Garcia papaya), capulín,* and *guaje (Leucaena esculenta);* a great number of medium-height trees and shrubs, including banana, peach, avocado, pomegranate, several types of citrus, *colorín (Erythrina americana), chirimoya (Annona cherimola),* and kumquat (*Fortunella*

[18]Edgar Anderson, *Plants, Man, and Life;* Clarissa Kimber, "Dooryard Gardens of Martinique"; Clarissa Kimber, "Spatial Patterning in the Dooryard Gardens of Puerto Rico"; Philip L. Wagner, *Nicoya: A Cultural Geography.*

Figure 12-4. Dooryard garden with multiple levels of productive plants.

spp.); and lower layers of high and low field crops such as maize, shrub beans, red and husk tomatoes, chile, and squash, interspersed with flowers and medicinal and cooking herbs. Economic vines including beans and chayote twined into otherwise unoccupied spaces. The whole garden was fenced with additional productive and ornamental plants, such as *colorín,* the bamboo-like *carrizo (Arundo donax),* and *jacaranda (Jacaranda acutifolia).* Finally, as if this floristic exuberance were not enough, chickens and turkeys patrolled between plants and contributed in their way to this productive, three-dimensional space (Figure 12-4).

Advantages

There are advantages to intercalating other than intensive use of space and greater total productivity. Multiple cropping reduces erosion by maintaining nearly a continuous plant cover over the soil for most of the year. Tall plants provide support for climbing vine crops, and leguminous plants replace soil nitrogen to the benefit of the whole crop area. Intercropped fields are less subject to infestation by above-ground pathogens and pests.[19] And diverse crops, especially if of different growth habits, replicate the structure and diversity of natural stands and create desirable, sheltered microenvironments.

[19]Miguel A. Altieri, *Agroecology.*

These features, long appreciated by traditional farmers, are attracting the attention of modern scientists. For example, data from Africa show reduced erosion and increased total yields from mixed cropping. Recent research in Costa Rica indicates that intercropping produces more usable products but fewer weeds and airborne diseases.[20]

Despite these advantages, most farmers would probably not use intensive techniques were space not a limiting factor. They feel that although total yields are higher, the quality of individual crops is not as high as it is when crops are planted separately. In addition, they consider intercropped fields more laborious than single-crop fields. Intercropped fields are difficult to work with animal-drawn equipment, and opportunities for rapid, uniform movements even in hand operations are reduced. The findings in Costa Rica generally bear out these impressions. Anderson and Norman, however, argue that mixed gardens and multiple cropping make efficient use of cultivators' time.[21] Additional research would clarify this critical issue. Meanwhile, most farmers I have consulted view intercropping as a response to space scarcity and would prefer to single-crop were sufficient land available.

Scheduling

Crop scheduling, or minimizing the time a crop is in the field, may be as important as crop spacing for increasing production per unit of area. Together, the two techniques form a powerful system for intensifying the use of productive farm space.[22] Simple forms of scheduling, such as crop rotation, are practiced in Middle America wherever growing seasons permit two or more crop cycles each year. Examples of complex

[20]D. J. Greenland, "Bringing the Green Revolution to the Shifting Cultivator"; Ruthenberg, *Farming Systems in the Tropics,* pp. 39–42.

[21]Centro Agronómico Tropical de Investigación y Enseñanza, "Small Farm Cropping Systems." Anderson, *Plants, Man, and Life;* Norman, "Rationalizing Mixed Cropping." Exceptions to this attitude are found where factors other than production are important. Thus, interplanting to control erosion is common even where land is abundant. The convenient location of dooryard gardens and cultivation by incidental labor distort the arguments for labor efficiency.

[22]Procedures for international crop reporting are based on the one-season, space-abundant agriculture common in midlatitude industrialized systems, and only with difficulty are adjusted to account for production from intercropped and multiple-cropped fields. For example, the Food and Agriculture Organization (*Report,* p. 115) mentions that "significant problems were encountered in the reporting of areas . . . where crops were grown simultaneously on the same land more than once during the agricultural year." See also Dalrymple, *Survey of Multiple Cropping,* pp. 5–8.

scheduling, where crops at different stages are grown together in the same field, are less common. Apparently, the ability to coordinate and integrate the planting, cultivating, and harvesting phases of several crops at different stages of growth is a skill shared by relatively few farmers. The examples are worth noting however. In the discussion here, general comments based on field observations are followed by a detailed review of seed beds, one of the important time- and space-conserving techniques used in Middle America.

Multiple Cropping

In its simplest form, scheduling involves nothing more than planting two or more crops in sequence so that productive space is used for as much of the agricultural year as possible. The main limiting factor is climate; there must be enough energy and moisture for enough time to permit two or more crop cycles during the growing season. But because the length of the growing season is also a function of growth periods and requirements of particular plants, in reality a combination of climate and crops establishes opportunities for successive crops in the same year.[23]

Growing seasons are limited by temperature or precipitation. Irrigation compensates for lack of rainfall. But even cold weather or frost can be partially offset by careful management of field microclimates. Thus, the amount of time a farmer can work his productive space depends on regional and local climate, crop characteristics, and mastery over such agricultural resources as water and temperature.

On most traditional farms, space must be allotted to a basic subsistence food plant. In Middle America, successive crops are planted only if the growing season is long enough to accommodate subsistence maize and one or more other crops. For example, one simple and popular highland succession consists of garlic, planted in late fall and harvested just before spring rains begin, followed by maize, planted in late spring and harvested in early autumn. Garlic is a cash crop; maize is for family consumption.

This example also illustrates the variable nature of growing seasons. Garlic thrives in the cool, dry winters of highland Middle America, whereas maize needs the warm, moist conditions of summer. Garlic has a relatively fixed growth period in a given environment. The various races of maize, however, have a wide range of growth periods, and most traditional farmers have experience with several types. Thus, in this case the two seasons—one cool and dry, the other warm and moist—are identified as growing seasons in relation to crop require-

[23] Jen-hu Chang, "A Critique of the Concept of Growing Season."

ments. Each is of variable length, also in relation to crop growth characteristics and limits.

Some farmers make an effort to speed up physiological processes. For example, many farmers on the Mesa Central of Mexico believe that removing maize tassels and leaves hastens maturation. Similarly the tips of inner tight leaf clusters *(meyolotes)* of *maguey* (*Agave* spp.) may be cropped off 20 to 50 cm from the ends each year to hasten opening of the leaves. Some *chinampa* farmers harvest still-green maize ears so their fields can be more quickly prepared for other crops. Drying is completed in houses or storerooms.

Substituting shorter-term crops, such as lettuce or radishes, for relatively long-term garlic, or shortening crop field time by planting seedlings from nursery beds might permit a three-crop cycle. But the real opportunities for successive crop cycles occur when subsistence maize is removed from crop sequences. Liberated from long-term, space-consuming maize, plots devoted exclusively to commercial crops can be scheduled through three and even four cycles each year.

Thus, one step in the intensification process is specialization of farm space. Farmers identify some plots for commercial production and others for subsistence. Because even the most commercially oriented traditional farmers continue to rely on home-produced food, maize production must continue. But instead of being alternated with commercial crops, the staple grain is relegated to less intensively managed, *temporal* fields, often in hilly areas, leaving the prime, intensively managed plots exclusively in commercial production. Some plots in Oaxaca, Mexico, are alternated between intensive vegetable cultivation and *temporal* maize farming, possibly to avoid salinization.[24] Although the spatial separation is temporary, it is no less real; while under intensive cultivation and irrigation, the plots are devoted exclusively to commercial vegetables, herbs, and flowers.

Phased Planting

But scheduling may also assume more complicated forms, as when crops with different growth and maturation periods are planted together or at staggered planting dates so that crops in the same field are at different stages. By this means, it is possible to continuously use not only horizontal but also vertical spaces that would otherwise go unoccupied. Possibilities for different crop combinations and schedules are almost infinite. An irrigated, 0.2-ha subsistence plot observed near Zacatelco, Tlaxcala, is an example. Visible in Figure 12-5 are seven

[24]Anne V. T. Kirkby, *The Use of Land and Water Resources in the Past and Present Valley of Oaxaca, Mexico,* p. 42.

Figure 12-5. Scheduled field, Tlaxcala, Mexico. Second-phase maize and bean seedlings appear beneath mature broad beans and associated plants.

different crop plants in different stages of growth: broad beans *(haba, Vicia faba),* climbing and shrub beans (*frijol enredador, frijol de mata;* both *Phaseolis vulgaris*), maize *(maíz, Zea mays),* chick-peas *(garbanzo, Cicer arietinum),* squash (*calabaza, Cucurbita* spp.), and peas *(arveja, chícharo; Pisum sativum).*

The first phase contains chick-peas, broad beans, and peas, all of which are mature and are being harvested. When the last of these are picked, the plants are removed (and used for fodder), exposing second-phase maize, bean, and squash seedlings. These were originally planted in furrows between rows of broad beans, as shown in Figure 12-6. In the second phase, rows are cross-plowed so that maize and associated plants are on ridges. Both crop phases contain several plants with different growth rates. Thus, this small plot will provide a nearly continuous stream of food over a relatively long growing period. Not only are there two distinct crop complexes but each is almost a structural duplicate of the other. Both contain emergent support plants (broad beans, maize), twining associates (peas, *enredador* beans), and intercalated shrubs (chick-peas, *mata* beans). Only the ground-hugging *calabaza* of the second phase has no morphological equivalent in the first.

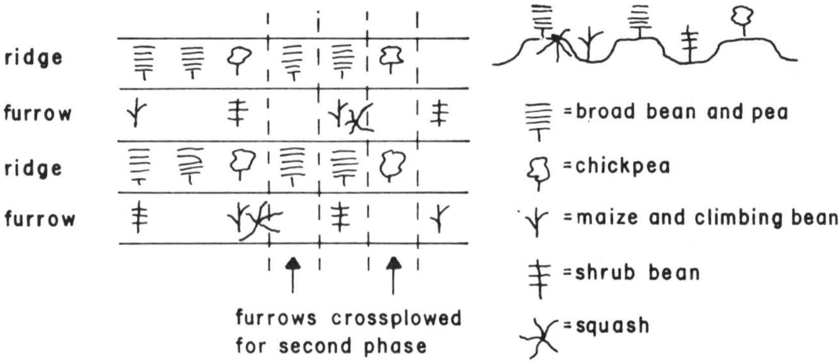

Figure 12-6. Multiple-cropped, two-phase field, Zacatelco, Tlaxcala, Mexico.

By incorporating quick-maturing vegetables, farmers increase the possibilities for scheduling enormously. For example, on commercial *tablones* in Quezaltenango, Guatemala, radishes with a maturation period of 30 to 35 days are planted with lettuce, which has an expected field time of 2.5 months. Some 20 to 30 days later, cauliflowers (2.5 to 3 months) are intercalated. Radishes are harvested before the lettuce grows to occupy much of the near-surface space. The taller cauliflower matures after the lettuce has been picked. Thus, although horizontal space is continuously occupied, competition for vertical space is minimized. In several vegetable regions of Middle America scheduling and multiple cropping are carried out during these separate crop phases. Farmers boast that if they harvest on Friday, they prepare the fields on Saturday and plant again on Monday.

Such tight scheduling and intensive cultivation require commensurate control of other agricultural resources. In fact, at some point the concept of natural resources slips away to be replaced by cultural resources because original characteristics have become so modified. Thus, field topography is more a product of farmers' decisions than of native configuration, soils are best described as anthrosols, irrigation substitutes for rain if not in quantity at least in reliability, and even field microclimates are partially controlled.

Longer-term scheduling opportunities arise when crop mixes include perennial shrub or tree crops. Thus, maize, wheat, beans, or peas are commonly planted in the open ground between rows of avocado, citrus, or mango seedlings in new orchards. Only after several years will spreading canopies reduce surface energy levels below those required for crop growth and development (Figure 12-7).

Figure 12-7. Maize interplanted with citrus, Veracruz, Mexico.

In dooryard gardens, complex spatial arrangements are paralleled by equally complex scheduling. Planting takes place continuously, and harvesting of seeds, stalks, leaves, roots, tubers, flowers, and fruits is a never-ending process. Until a more precise timetable of operations is developed, Anderson's description of a Guatemalan garden will serve:

> The garden was in continuous production but was taking only a little effort at any one time: a few weeds pulled when one came down to pick the squashes, corn and bean plants dug in between the rows when the last of the climbing beans were picked, and a new crop of something else planted above them a few weeks later.[25]

Seed Beds

If seeds are planted directly in fields at intervals appropriate to mature plants, then for weeks tiny sprouts and seedlings utilize but a fraction of the productive space. Mass seeding and later thinning, as is common with some vegetables, increases ultimate plant density by reducing space lost to unviable seeds or seedling death. But it does not solve the problem of underutilization of field space during germination and seedling periods.

[25]Anderson, *Plants, Man, and Life,* p. 141.

Figure 12-8. Tiny seed beds adjacent to fields.

One solution is to set aside special planting areas or seed beds (Spanish from the Latin, *sementeras;* Spanish from the Arabic, *almácigas;* Nahuatl *centemilli, tlachtli*) and, by so doing, to substitute an early microspace period for part of total field, or macrospace, time. Although the tiny seed or nursery beds take but a fraction of total field area, they provide all the room necessary for the first few weeks or months of plant life. In fact, because they are so small, they are easy to tend with special care, and it is here that some of the most advanced traditional resource-management techniques are found (Figure 12-8).

But the primary advantage of seed beds in space-scarce systems is conservation of productive field space. The results are significant. For example, in Oaxaca, Mexico, tomatoes are commonly started in seed beds around the first of November, transplanted to fields in mid-December, and harvested during a two-month period from April to June. Thus, the field period of tomatoes is reduced by 20 to 25 percent, during which time the space can be used by other crops. Even greater space savings are realized with onions, which pass 1.5 to 2 months *en almáciga* followed by an additional 3 months in the field, for a total saving of 30 to 40 percent of field time. Other annuals often started in seed beds are celery, cabbage, beets, lettuce, swiss chard, cauliflower, leeks, and rutabaga. Although grains are generally sown or planted directly in fields, maize is still occasionally started in seed beds in the extraordinarily crowded *chinampa* region of Mexico.[26]

Even when space is abundant, savings in field space and ease of caring for seedlings prompt the use of seed beds for slow-growing plants. For example, tree or shrub crops such as coffee, citrus, and cacao are commonly started in seed or nursery beds. The concern here is to commit field space for the long periods of plant life only to those plants with the best prospects for growth and production. In seed beds seedlings can be observed continuously, and their environment can be managed to ensure optimum amounts of fertilizers, water, and light during early growth stages. At transplanting time, field space is committed to only the most promising. The process well illustrates the interrelated nature of time- and space-saving management techniques. Clearly, the longer the seedling stage, the more desirable seed-bed planting is.

As with so many traditional resource-management techniques, however, the cost of seed beds is labor. Time and energy are required to construct and maintain carefully tended seed beds and to transplant seedlings, and returns must justify the expenses. Two examples from central Mexico illustrate these characteristics.

On the Mesa Central of Mexico, chile (*Capsicum* spp.) is first densely sown in long narrow *almácigas* in November or December. Seeds are sown on carefully prepared and manured seed-bed surfaces, then sprinkled with soil and covered with a layer of branches weighed down with earth to insulate them against frosts. After a month or so the overlying cover is removed and the branches are formed into arbors, which continued to offer protection against frost, hail, and birds. At this stage, counts in randomly selected 20-cm-×-20-cm squares revealed seedling densities from 0 to 200, with an average of about 60 plants/400 cm^2, or about 1,500/m^2.

Chile seedlings continue development for another three to four months. Then about mid-April they are transplanted to open fields in bunches of three to six (the average is four) and, depending on the type, are spaced 30 to 50 cm apart, usually in staggered rows. Staggering maintains standard intervals between plants and yet increases plant spacing by about 13 percent over opposed planting. At 30-cm spacing, each cluster of transplanted seedlings takes up about 0.08 m^2, and about

[26]Robert C. West and Pedro Armillas, "Las Chinampas de México."

0.22 m^2 at 50-cm spacing. Because seed-bed density was 1,500/m^2 and each transplanted cluster contains an average of four seedlings, expansion factors from seed beds to fields range from 1:30 to 1:75. That is, one square meter of chile seed bed supplies seedlings for 30 to 75 m^2 of field space.[27]

Chile is harvested in September and October. Thus of a 10- to 11-month total growing season, 40 to 50 percent may be spent in seed beds. Or, put another way, the chile crop occupies only 2 to 4 percent of productive field space for almost half the total growth period. The benefits are substantial. In frost-free regions, the 4 or 5 months seedlings spend in *almácigas* are enough to produce another crop in the main field, such as green maize for roasting ears *(elotes)* or vegetables.

Although some features of seed beds in the *chinampa* district have already been examined, this unusual procedure merits a brief review. Small sections of *chinampa* plot surfaces are carefully worked and leveled with hoes, and ridged around with low borders 3 to 10 cm high. Rich, semiliquid muck scooped from canal bottoms is laced with chemical fertilizers and poured into seed beds to a depth of 3 to 5 cm for most vegetables, 8 to 10 cm for maize, and even deeper for some flower stocks such as roses. The muck settles and solidifies within a few hours; it is then cut with knives into squares with sides about equal to muck depth. Holes poked in the center of each cube with fingers or sticks receive seeds that are then covered with a light cap of pulverized manure and soil and carefully watered with sprinkler cans. At transplanting, the entire root cube *(chapín)* is transferred to the main plot—a process that also effectively distributes the rich muck and fertilizer over the plot surface.[28]

This extraordinarily meticulous procedure not only assures high seedling survival rates but also achieves the time and space savings associated with conventional seed beds. For example, six to eight spinach seeds are planted in root cubes 3 cm on a side. Some 15 to 20 days later the *chapines* are transferred to main plots at a spacing of 10 to 12 cm for an expansion ratio of 1:11 to 1:16. Harvest takes place 1.5 to 2 months later. Thus, the crop occupies only 6 to 9 percent of the *chinampa* surface for 25 to 45 percent of its life.

Chinampa land is so scarce and so valuable that scheduling is a major if not dominant factor in land use. *Chinamperos* (*chinampa* farmers) are reluctant to plant any crop that will occupy a field for more

[27]The close relationship between chile cultivation and seed beds in the minds of farmers is manifested during transplanting, when field hands call to helpers for fresh supplies of seedlings with the cry *"almásiga!"*

[28]An almost identical method of transplanting in soil blocks was developed in Holland and is currently being promoted along with simple equipment to form the blocks. David Tresemer, *Transplants in Soil Blocks.*

than a few months; they favor such rapid-turnover crops as lettuce, radishes, and spinach. Some slower-growing crops, such as cauliflower, command sufficiently high prices to warrant *chinampa* cultivation. But slow-maturing, low-value crops, such as potatoes, are not popular.

Summary

By incorporating crops that occupy different horizontal and vertical spaces, multiple cropping makes efficient use of field space. When multiple cropping is coupled with scheduling, the use of spaces becomes more continuous. A further step is to set aside separate areas where crops pass germination and early seedling stages under special care and where they occupy only a fraction of valuable field space.

For exceptionally slow-growing crops, such as orchard trees and shrubs, seed-bed costs of time and energy are justified by long plant life even in areas with abundant space. In areas where space is scarce, transplanting short-cycle crops is common. Integrating seedlings from seed beds into intercalated and phase-planted fields permits exceptionally tight scheduling and further reduces losses of productive field space.

Close management of field space and schedules increases productivity per unit area. Increased productivity comes at the price of increased labor, however. The highest crop densities are found in farming systems that use the least fossil fuel or even supplements of animal energy. Intensive crop scheduling requires close control of soils, water, and other elements of the crop environment. And, although seed beds save substantial amounts of field time, they involve arduous transplanting, an operation that has become a symbol of hard field labor. Thus, space- and time-conserving methods are employed where space is physically, ecologically, or economically scarce, and appear to be a response to that scarcity.

13

Traditional Resource Management in the Future

The main objectives of this book were to examine the possibilities for farming when certain limitations of energy, technology, and scale apply and to develop some principles of traditional management. Observations came mostly from areas of intensive cultivation and dense rural populations, where traditional methods are most advanced. Some of the procedures described are atypical in that they represent state-of-the-art rather than run-of-the-mill technology. Nevertheless, they illustrate what can be achieved within the imposed constraints.

Although dispelling myths about traditional agriculture was not a major objective of this study, several have become casualties of the data. For example, although traditional farms have little capital equipment, the labor invested in structures for managing resources may be substantial. Furthermore, the concept of invested labor confounds the idea of investing capital to replace labor. When the investment medium is labor itself, labor may be expended to save labor. Such investments of labor were particularly striking in the *galerías filtrantes* and silting *pantles* of Tehuacán, but there may be similar cases in many traditional management systems.

The impression that nonindustrial technology makes traditional agriculture a purely ecological operation and not a social and economic one has been tarnished. That many of these practices are old[1] suggests

[1]Middle American agricultural systems are remarkably persistent. As far as I know, all resource-management techniques reported in archaeological or

that they are in some sort of ecological equilibrium. But they are also eminently practical techniques developed under specific economic and social conditions, and they will likely be abandoned if these conditions change.

Even the notion of appropriate technology is badly represented. "Appropriate technology . . . has come to mean technology which is decentralized, small in scale, labor intensive, amenable to mastery and maintenance by local people, and harmonious with local cultural and environmental conditions."[2] Although constraints of energy and scale exclude use of the large machines of the industrialized world, complex systems, such as the silting *pantles* of Tehuacán or the *chinampas* of Lake Chalco, seem poor examples of the simple equipment that is supposedly harmonious with village life. The problem again lies in narrow definitions. If technology includes not only equipment but also social organization and procedures, then the traditional world can offer its own examples of complex and specialized technologies that would be quite inappropriate for other societies.

Finally there is the traditional farmer himself, who emerges in sharp contrast to some cultural and technological myths. Any lingering images of the lazy, dull, noneconomic peasant farmer must surely have vanished in the sun-baked *cántaro* fields or the tightly organized *galería* societies. Given any sort of reasonable reward, traditional farmers respond much like their industrialized counterparts: they identify goals, minimize costs (most of which are labor), and maximize gains by careful use of available resources. Their management methods are also not "traditional" in the sense of being stubbornly preserved anachronisms of dubious or even negative value.[3] Instead, they are agronomically and hydrologically sound, based on a thorough understanding of local conditions and plant requirements, and deserve close examination by the scientific community.

The idea that the methods of traditional agricultural and resource management merit serious consideration is not new. Traditional techniques from the simplest to the most complex attracted the attention of foreign field workers who saw their potential and recommended wider use long before advocating "appropriate" technology was

---

historical sources are still practiced in contemporary Middle America. The only possible exceptions are the drained-field systems of the ancient Maya, for which there is little information on construction and operation. Even these, however, likely resembled drainage techniques used elsewhere in the region.

[2]Rashid Shaikh, "Commentary."

[3]"Tradition . . . [therefore] does not represent a stultified past, but a living past that is interpreted in the light of a significant present." Kenneth Zaretske, "The Idea of Tradition," p. 91.

fashionable.[4] More recently, international research centers, private organizations, and development agencies have supported programs that incorporate traditional methods.[5] The recognition is encouraging, although perhaps even more encouraging are the traditional systems that are expanding spontaneously in response to fresh market opportunities. But before advocacy, the great need is to understand the workings of traditional technologies and the possible roles they might play in an increasingly crowded and demanding world.

These are difficult times for agriculture. Demand for farm products grows as population and consumption levels increase. But in imperfect exchange systems demands are ineffectively expressed, and surpluses accumulate in one region at the same time that desperate needs go unmet in others. Despite much-proclaimed global and regional links and interdependence, national and even regional isolation are strong arguments for efficient local production. At the same time, resources, especially land but also water and energy, are limited and increasingly costly. The search for solutions has uncovered many new crops and methods, some of which show great promise but many of which are as yet unproven. The solutions of the past however are certain: higher yielding varieties and better resource management.

Unfortunately, the discussion becomes polarized at this point because of conflicts over the type of management that is needed. Critics of traditional methods point to apparently low productivities, especially of labor, and argue that the solution lies in accelerated adoption of Western technology. This position is understandably popular in technologically advanced countries.[6] Yet industrialized agriculture itself is not without detractors who disparage the heavy use of chemicals and fossil fuels, adverse environmental effects, and large-scale labor displacements, and who advocate a return to presumably ecologically and socially harmonious traditional practices.[7] It is ironic that the two

[4]Most if not all major traditional agricultural techniques have found advocates among foreign technical experts who have come to appreciate their merits or at least their potentials.

[5]Derek Byerlee, Larry Harrington, and Donald L. Winkeleman, "Farming Systems Research"; John L. Dillon, Donald L. Plucknett, and Guy J. Vallaeys, *Farming Systems Research at the International Agricultural Research Centers.* Similarly, countries are now beginning to recognize the possibilities offered by their traditional technology. Grupo de Investigaciones Agrarias, *En el Centro de Documentación de Tecnología Campesina Tenemos.*

[6]For a concise review of development models and induced technical innovations, see Vernon W. Ruttan, *Agricultural Research Policy,* especially chap. 2, "Technical Change and Agricultural Development."

[7]Miguel A. Altieri, Deborah K. Letourneau, and James R. Davis, "Developing Sustainable Agroecosystems"; J. Bandyopadhyay and V. Shiva, "Alternatives for India"; Donald Q. Innis, "The Future of Traditional Agriculture"; Warren Johnson, *Muddling Toward Frugality.*

archetypal management approaches, which together account for most of the present world's agricultural production, have both been declared untenable by their more vociferous critics.

Both positions suffer from a distressing lack of supportive evidence. The record of technology transfers from industrial to traditional societies is not good.[8] On the other hand, data on the productivity of traditional agriculture is meager, and even less is known about how these systems might respond to changed conditions in the future. With contradictory evidence on both sides, the arguments often seem more defenses of economic or academic positions than dispassionate searches for solutions.

But is it necessary to choose? Are there only two alternatives? For convenience I established the criteria of energy, technology, and scale to differentiate between management systems. But the associated labels—small and large scale, traditional and modern, industrial and nonindustrial—that I introduced to clarify the alternatives may instead confuse, especially if they are applied to agricultural production rather than resource management. Clearly there are no overwhelming advantages to any one management system in the many social, economic, and environmental situations where farming takes place.[9] Thus, the designations *traditional* or *modern* hardly provide a basis for evaluation. Instead, actual results must be compared.

But what results? Unfortunately, many of the conflicting claims about farming systems are based only on production per unit of land or labor. Here, the strict criteria applicable to modern, specialized, single-crop systems fail to cover the multiple uses of land and labor in traditional agriculture.[10] Even on a straight production basis, the situation is not clear. Although traditional agriculture generally fares poorly in comparison with modern systems, some of the world's highest yields come from traditionally managed small farms.[11] The fact that the traditional world is also a nonstatistical world creates further disadvan-

[8]M. Taghi Farvar and John P. Milton, *The Careless Technology.*

[9]George MacPherson and Dudley Jackson ("Village Technology for Rural Development") develop a convincing argument for not just one but a variety of technologies, each suited to the constraints of peasant farming.

[10]Gene C. Wilken, "The Fallacy of Labour/Land Comparisons of Agricultural Productivity." Some of the difficulties of developing uniform data from differing national field and statistical information are suggested in Food and Agriculture Organization, "Crop Area."

[11]"The general conclusion . . . is that the small-farm sector makes better use of its available land than does the large-farm sector, largely through applying higher levels of labor inputs (family labor) per unit of land." R. Albert Berry and William R. Cline, *Agrarian Structure and Productivity in Developing Countries,* pp. 131-32. See also World Bank, *World Development Report 1980,* p. 42.

tages for small-scale farmers. Gross harvests are often underreported and home consumption, local exchange, and postharvest losses may escape accounting entirely.[12] More and better data than are now available would do much to alleviate the confusion.

But perhaps before data we need objectives. In agriculture there are many: an acceptable standard of living, increased income, and security head the list but do not exhaust the reasons for investing time and effort in resource management. Only a few studies have attempted to identify the objectives and then to analyze the efficiency with which a particular management procedure achieves these objectives.[13] This analytical approach is common at national levels, where crops, technologies, and even land tenure are periodically reexamined to ascertain whether such policy goals as self-sufficiency or export earnings are being met. It is less common at the level of the individual farm, where family goals and alternatives are seldom clearly stated and where options for changing crops or management methods are limited. But ends are no less linked to means at local levels than they are at national levels, and a fair case could be made for differentiating farming systems on the basis of goals and objectives rather than on the customary criteria of crop mixes, technology, and farm size.[14]

In addition to goals, the relationships between investments of labor and returns have not been examined here. Resource management enhances plant environments and thus crop growth and development. By inference, the objective is increased production; the test of efficiency is whether management costs are commensurately rewarded by yields; the assumption is that somehow increased production will help farmers attain their goals. By and large the farmers whose techniques are reviewed here share this attitude toward production and the assumption regarding rewards.

But problems arise in turning increased production into commensurate rewards. Usually this final step is accomplished through exchange or market systems. But access to markets may be obstructed, either physically by distance or poor roads or socially by groups economically

[12]Kenneth L. Harris and Carl J. Lindblad, *Postharvest Grain Assessment Methods;* National Academy of Sciences, *Postharvest Food Losses in Developing Countries.*

[13]Peggy F. Bartlett *(What Shall We Grow?)* reviews the making of decisions to satisfy family needs.

[14]"Before we choose our tools and techniques, we must choose our dreams and values, for some technologies serve them, while others make them unobtainable." Tom Bender, "Appropriate Technology." Clifton R. Wharton, Jr. ("Research Priorities on Subsistence Agriculture"), argues that values, motives, attitudes, and aspirations should be included in research on subsistence agriculture. Although designations such as *subsistence* and *commercial* suggest goal orientations, few of the studies that use the terms analyze the goals.

or ethnically distinct from the producers. In such cases physical and social distances between production and marketing systems uncouple the links between costs and benefits and thus between means for achieving goals and the realization of goals.[15]

An analysis of goals, means, and effectiveness would carry us well beyond a review of traditional technology and into the realm of development.[16] It is enough here to set forth some positions that are supported or suggested by this study. I briefly consider five areas: production, conservation, employment, income, and technological change.

Production

Surveys and individual studies have revealed no consistent evidence that yields per unit of area vary according to farm size. In fact, some of the highest per-hectare yields come from intensively worked plots where multiple cropping and intercropping and careful resource management are practiced in traditional fashion. Clearly, some plants respond to the close attention and to the techniques possible only in small-plot farming, whereas others seem relatively unaffected or do equally well under extensive management. Rather than generalizing about productivity under traditional or modern management approaches, analysts should examine individual factors such as technology and scale with reference to specific crops.

There are few technical barriers to increasing production on traditional farms. As amply demonstrated here, the basic inputs of fertilizer, water, and seed[17] are neutral with respect to scale, or may even

[15]For examples of the complex social and economic circumstances under which traditional farmers produce and sell their crops, see Peggy F. Bartlett, *Agricultural Choice and Change;* Sutti R. Ortiz, *Uncertainties in Peasant Farming.* Amir U. Khan ("Appropriate Technologies") stresses the need to consider market factors before selecting technologies.

[16]Cynthia Hewitt de Alcantara *(Modernizing Mexican Agriculture,* p. xiii) provides a clear distinction between modernization and development: "Modernization simply suggests a process of reorganization within groups or societies engaged (directly or indirectly, willingly or unwillingly) in an effort to increase control over their physical environment through recourse to new tools and methods, most particularly those associated during the last two hundred years with the industrial revolution in Europe and America. But development, if it is to have any meaning at all, is an ethical concept: it implies an ethical judgment of the process of change."

[17]In this case the term *fertilizer* includes organic as well as inorganic amendments; *seed* includes local varieties as well as hybrids that may be selected for particular fields or seasons.

be easier to apply in small rather than large operations. In some instances mechanized equipment can achieve better results than nonmechanized procedures, as is the case when deep plowing with tractors eliminates subsurface pans, or when powered harvesters improve scheduling. But in general most yield-increasing technology can be applied effectively to small plots with the tools available to small-plot farmers.[18]

Conservation

Some traditional systems have enviable resource conservation records; others do not. But those that have maintained or enhanced the resource base over time merit examination. These successful experiments by generations of farmers in particular places constitute valuable lessons that should not be ignored or lost.

But before our enthusiasms for centuries-old systems outrun the record, we should note the failures. It is also condescending to view traditional farmers as sagacious husbandmen, imbued with infallible folk wisdom, in mystic harmony with the environment. In fact, they share with their industrialized counterparts the human propensities to respond to short-term opportunities while disregarding long-term costs, to misinterpret the durability and flexibility of agroecosystems, and generally to err.[19] The many past and present examples of depleted resources and degraded landscapes attest to the potential for traditional mismanagement. But failure is part of the agricultural experience and has as much to teach, in its own way, as success. This is especially true in systems under stress, where old responses to new conditions have resulted in management failures.

Employment

Reliance on human and animal power is to Western eyes one of the most onerous aspects of traditional agriculture. Caught in the endless

[18]"Changes in the size of farm do not seem to be a major prerequisite in the adoption of yield-increasing technology; size of farm does not create cost barriers nor are increases in land-man ratio required for such technological changes." Don Kanel, "Size of Farm and Economic Development." See also Kenneth L. Bachman and Raymond P. Christensen, "The Economics of Farm Size."

[19]George A. Collier (*Fields of the Tzotzil,* p. 109) cautions against the image of "primitive" societies living in finely balanced harmony with nature; in fact, examples abound of populations in obvious disequilibrium with their environments.

round of daily and seasonal tasks, the farmers seem to live in drudgery from which anyone would welcome escape. There is no question that a full day irrigating *a cántaro* or forming terraces *con azadón* is arduous and repetitious. But is labor-intensive farm work truly more distasteful or dehumanizing than the routine tasks of offices or factories? The evidence suggests that farmers are forced from their land and rural life more often by social and economic circumstances than by the work itself.

In fact, labor-intensive methods may be attractive to labor-abundant societies, where alternative employment opportunities are scarce. The population-supporting qualities of small-farm sectors are well known, as are the disadvantages of excessive migration from rural to urban areas.[20] Less well known are the means by which dense rural populations can be usefully employed. Here, traditional technology has some clear advantages over the labor-saving technology of industrialized systems.[21]

In addition, lack of dependence on fossil fuels may be appealing not only to the countries that must purchase fuels but also to the farmers themselves. Fuels and maintenance for traditional tools and equipment are procured locally, usually at lower cost and with greater reliability than are fuels and maintenance for externally produced powered equipment. In societies with imperfect distribution systems the relatively higher labor requirements associated with local technology may be an acceptable price to pay for the use of traditional equipment.

Income

The real questions are whether traditional farmers can drive their small plots to high levels of productivity and, if they can, whether they will realize a satisfactory return for their efforts. As we have seen, the answer to the first part depends on the crops and resources available for management. The second part of the question is more difficult because it involves economic and social institutions that are outside the farmers' limited area of operation and control. Although not reviewed here, market accessibility should be considered as essential as any resource or production technology. The expectation of reasonable returns on

[20]Wyn F. Owen, *Two Rural Sectors.*

[21]Hoyt Alverson, "Feeding Africa"; Harry T. Oshima, "The Time to Change to Labour-Intensive Policies Is Now"; Stephen M. Freedman, "Human Labor as an Energy Source for Rice Production in the Developing World"; Bernardo M. Villegas, "For Appropriate Technology That Uses the Resources of the Third World by Substituting Labour for Capital"; G. M. Ward, T. M. Sutherland, and J. M. Sutherland, "Animals as an Energy Source in Third World Agriculture."

invested labor is a powerful incentive and may be the most important reason why some traditional farmers diligently practice careful resource management while others do not.

Technological Change

Traditional technology has little to attract Western engineers and businesses. Yet because it represents the means by which the labor of millions of farmers and their animals is used and millions of hectares of productive land are managed, even small improvements would be significant.[22] Transferring existing tools and techniques between societies that are environmentally, technically, and socially comparable would provide some advantages over introductions of modern equipment. Capital requirements are low and adoptions require little restructuring of traditional procedures or institutions. There is no reason why development efforts should not diffuse traditional technology along with methods and tools from technically advanced societies.

But it is not enough to extol the virtues of traditional ways or to naively suggest their wider use. The traditional world contains a truly vast array of methods for managing crops and resources under a variety of environmental and social conditions. Except in the crucible of experience they remain largely untested not only for ecological soundness but also for effectiveness in realizing individual and national goals. Even the procedures by which traditional technology could be examined and eventually incorporated into the general body of agricultural knowledge remain vague.[23]

If nothing else, this study of a few resource-management practices in one small region has perhaps demonstrated something of the nature and complexity of traditional agricultural technology. Clearly we are not dealing with only a few tools and techniques. In fact, the tools are usually simple and multipurpose and are the lesser part; most of tra-

[22]H. J. Hopfen and E. Biesalski, *Small Farm Implements,* p. 72; S. R. Gliessman, R. Garcia E., and M. Amador A., "The Ecological Basis for the Application of Traditional Agricultural Technology in the Management of Tropical Agro-Ecosystems"; MacPherson and Jackson, "Village Technology"; Max F. Millikan and David H. Hapgood, *No Easy Harvest,* pp. 44–48.

[23]For recent discussions of the potentials of traditional agriculture, see David W. Brokensha, D. M. Warren, and Oswald Werner, eds., *Indigenous Knowledge Systems and Development;* D. J. Greenland, "Bringing the Green Revolution to the Shifting Cultivator"; Uma Lele, "Rural Africa." Procedures for developing benefit/cost analyses for some small-farm operations are discussed in Richard K. Perrin et al., *From Agronomic Data to Farmer Recommendations.*

ditional technology consists of understanding local conditions and ways of managing resources by using local energy and materials. Most of this technology remains unknown outside local areas. There is much to be learned about how these practices work. Fortunately, among the millions of traditional fields and farmers, there are many classrooms, many teachers.

Bibliography

Aeppli, Hans, "Aclaraciones acerca del Diseño del Mapa del Suelos de la Cuenca de Puebla-Tlaxcala." *Comunicaciones* (Fundación Alemana para la Investigación Científica) 6(1972):7-13.

Aguirre Beltran, Gonzalo, *Regiones de Refugio: El Desarrollo de la Comunidad y el Progreso Dominical en Mestizo América.* Ediciones Especiales no. 46, Mexico City: Instituto Indigenista Interamericano, 1967.

Ahmad, Chaudhry Nuruddin, *Plant Uptake of Water from a Water Table.* Water Management Technical Report 41. Fort Collins, Colo.: Consortium for International Development, 1975.

Alemi, M. H., "Distribution of Water and Salt in Soil Under Trickle and Pot Irrigation Regimes." *Agricultural Water Management* 3(1980/1981):195-203.

Allen, William, *The African Husbandman.* Edinburgh: Oliver & Boyd, 1965.

Alonso de San Juan [supposed author], *Relación Breve y Verdadera de Algunas Cosas de las Muchas que Sucedieron al Padre Fray Alonso Ponce en las Provincias de la Nueva España.* 2 vols. Madrid: Imprenta de la Viuda de Calero, 1873.

Alsina Grau, Luis, *Calendario del Horticultor.* Barcelona: Editorial Sintes, 1973.

Altieri, Miguel A., *Agroecology: The Scientific Basis of Alternative*

Agriculture. Berkeley: Division of Biological Control, University of California, 1983.

Altieri, Miguel A., Deborah K. Letourneau, and James R. Davis, "Developing Sustainable Agroecosystems." *BioScience* 33(1983):45–49.

Alverson, Hoyt, "Feeding Africa: A Dissent from Development." *Issue* 8(Winter 1978):20–22.

American Society of Agronomy, *Multiple Cropping*. Madison, Wis.: American Society of Agronomy, 1976.

Anderson, Edgar, *Plants, Man, and Life*. Berkeley and Los Angeles: University of California Press, 1967.

Armendaris, Eduardo, *Estudios sobre las Aquas de Tehaucán*. Mexico City: Oficina Tip. de la Secretaría de Fomento, 1902.

Armillas, Pedro, "Notas sobre Sistemas de Cultivos en Mesoamérica: Cultivos de Riego y Humedad en la Cuenca del Rio de las Balsas." *Anales del Instituto Nacional de Antropología e Historia* 3(1949):85–113.

Armillas, Pedro, "Land Use in Pre-Columbian America." In *A History of Land Use in Arid Lands*, ed. L. Dudley Stamp. UNESCO Arid Zone Research 17. Paris: United Nations Educational, Scientific, and Cultural Organization, 1961.

Armillas, Pedro, "Gardens on Swamps." *Science* 174(1971):653–61.

Arnold, Dean E., "Ethnomineralogy of Ticul, Yucatan Potters: Etics and Emics." *American Antiquity* 36(1971):20–40.

Arriola, Jorge Luís, *El Libro de las Geonimias de Guatemala*. Guatemala City: Editorial "José de Pineda Ibarra," 1973.

Ashton, E. H., "A Sociological Sketch of Sotho Diet." *Transactions of the Royal Society of South Africa* 27(1939):147–214.

Bachman, Kenneth L., and Raymond P. Christensen, "The Economics of Farm Size." In *Agricultural Development and Economic Growth*, eds. Herman M. Southworth and Bruce F. Johnson. Ithaca: Cornell University Press, 1967.

Bandyopadhyay, J., and V. Shiva, "Alternatives for India: Western or Indigenous Science." *Science for the People* 13(Mar.-Apr. 1981):22–28.

Barber, Edwin Atlee, *The Ceramic Collector's Glossary*. New York: Da Capo Press, 1967.

Bartlett, H. H., "Fire, Primitive Agriculture, and Grazing in the Tropics." In *Man's Role in Changing the Face of the Earth*, ed. William L. Thomas, Jr. Chicago: University of Chicago Press, 1956.

Bartlett, Peggy F., *What Shall We Grow? A Critical Survey of the Literature on Farmers' Decision Making*. Washington, D.C.: Bureau for Program and Policy Coordination, U.S. Agency for International Development, 1979.

Bartlett, Peggy F., *Agricultural Choice and Change: Decision Making in a Costa Rican Community*. New Brunswick, N.J.: Rutgers University Press, 1982.

Bear, Firman E., *Soils in Relation to Crop Growth*. New York: Reinhold, 1955.

Becerril Colín, Alfredo, "Proyecto Preliminar de la Planificación de la Cuenca del Valle de México." *Ingeniería Hidráulica en México* 15, no. 3(1961):71-82.

Bender, Tom, "Appropriate Technology." In *Rainbook*. Portland, Oregon: Schocken Books, 1977.

Berlin, Brent, Dennis E. Breedlove, and Peter H. Raven, "Folk Taxonomies and Biological Classification." *Science* 154(1966):273-75.

Berlin, Brent, Dennis E. Breedlove, and Peter H. Raven, "General Principles of Classification and Nomenclature in Folk Biology." *American Anthropologist* 76(1973):214-42.

Berlin, Brent, Dennis E. Breedlove, and Peter H. Raven, *Principles of Tzeltal Plant Classification*. New York: Academic Press, 1974.

Berry, R. Albert, and William R. Cline, *Agrarian Structure and Productivity in Developing Countries*. Baltimore: Johns Hopkins University Press, 1979.

Beyer, H. O., "The Origin and History of the Philippine Rice Terraces." *Proceedings of the Eighth Pacific Science Congress* 1(1955):387-98.

Black, J. N., "Energy Relations in Crop Production." *Annals of Applied Biology* 67(1971):272-78.

Blanc, M. L., H. Geslin, I. A. Holzberg, and B. Mason, *Protection Against Frost Damage*. Technical Note 51 (WMO 133). Geneva: World Meteorological Organization, 1963.

Blásquez L., Luis, "Hidrogeología de una Parte del valle de Tehuacán, Edo. de Puebla." *Anales del Instituto de Geología* (Universidad Nacional Autónoma de México), Vol. 12, *Estudios Hidrogeológicos*, 1957.

Blásquez L., Luis, and Raúl Lozano García, "Hidrogeología y Minerales No-metálicos del Estado de Tlaxcala." *Anales del Instituto de Geología* (Universidad Nacional Autónoma de México) 8(1946):i-iv, 1-100.

Boers, T. M., and J. Ben-Asher, "A Review of Rainwater Harvesting." *Agricultural Water Management* 5(1982):145-58.

Bonis, Samuel B., *Geología del Area de Quezaltenango, República de Guatemala*. Guatemala City: Instituto Geográfico Nacional, Ministerio de Comunicaciones y Obras Públicas, 1965.

Boserup, Ester, *The Conditions of Agricultural Growth*. Chicago: Aldine, 1965.

Bostwick, Don, *Studies in Yield Variability*. Montana State College

Bulletin 574. Bozeman: Montana Agricultural Experiment Station, 1963.

Brady, Nyle C., *The Nature and Properties of Soils.* 8th ed. New York: Macmillan, 1974.

Brambila, Miguel R., "The Tepetate Soils of Mexico." *Proceedings of the Sixth Pacific Science Congress* 4(1940):869-71.

Brokensha, David W., D. M. Warren, and Oswald Werner, eds., *Indigenous Knowledge Systems and Development.* Lenham, Md.: University Press of America, 1980.

Bromley, Daniel W., *Improving Irrigated Agriculture: Institutional Reform and the Small Farmer.* Staff Working Paper 531. Washington, D.C.: World Bank, 1982.

Brookfield, Harold C., "New Directions in the Study of Agricultural Systems in Tropical Areas." In *Evolution and Environment,* ed. Ellen T. Drake. New Haven, Conn.: Yale University Press, 1968.

Brookfield, Harold C., "Intensification and Disintensification in Pacific Agriculture." *Pacific Viewpoint* 13(1972):30-48.

Bryan, Kirk, "Flood-Water Farming." *The Geographical Review* 19(1929):444-56.

Burrows, W. C., "Characterization of Soil Temperature Distribution from Various Tillage-Induced Microreliefs." *Proceedings of the Soil Science Society of America* 27(1963):350-53.

Byerlee, Derek, Larry Harrington, and Donald L. Winkelmann, "Farming Systems Research: Issues in Research Strategy and Technology Design." *American Journal of Agricultural Economics* 64(1982):897-904.

Caborn, J. M., *Shelterbelts and Microclimates.* Forestry Commission Bulletin 29. Edinburgh: H.M. Stationery Office, 1957.

Carson, M. A., and M. J. Kirkby, *Hillslope Form and Process.* Cambridge: University Press, 1972.

Carter, William E., *New Lands and Old Traditions: Kekchi Cultivators in the Guatemalan Lowlands.* Latin American Monographs, 2d ser., no. 6. Gainesville: University of Florida Press, 1969.

Centro Agronómico Tropical de Investigación y Enseñanza, "Small Farm Cropping Systems" (titles vary). *Activities at Turrialba,* various issues, 1977-.

Centro Internacional de Mejoramiento de Maíz y Trigo, *The Puebla Project 1967-69.* Mexico City: Centro Internacional de Mejoramiento de Maíz y Trigo, n.d.

Chacón, J. C., and S. R. Gliessman, "Use of the 'Non-Weed' Concept in Traditional Tropical Agroecosystems of Southeastern Mexico." *Agro-Ecosystems* 8(1982):1-11.

Chang, Jen-hu, *Climate and Agriculture.* Chicago: Aldine, 1968.

Chang, Jen-hu, "A Critique of the Concept of the Growing Season." *The Professional Geographer* 23(1971):337-40.

Chisholm, Michael, *Rural Settlement and Land Use.* London: Hutchinson University Library, 1962.

Clark, Colin, and Margaret Haswell, *The Economics of Subsistence Agriculture.* 4th ed. London: Macmillan, 1970.

Cleek, Richard K., "The Infiltration Gallery in Mexico: Two Diffusions." *Anthropos* 68(1973):897-909.

Coe, Michael D., "The Chinampas of Mexico." *Scientific American* 211(July 1964):90-98.

Collier, George A., *Fields of the Tzotzil.* Austin: University of Texas Press, 1975.

Comisión de La Malinche (Mexico), *Realizaciones en la Montaña de La Malinche.* Mexico City: Secretaría de Agricultura y Ganadaría, 1963.

Conklin, Harold C., *Hanunóo Agriculture: A Report on an Integral System of Shifting Cultivation in the Philippines.* FAO Forestry Development Paper 12. Rome: Food and Agriculture Organization, 1957.

Conklin, Harold C., *Folk Classification.* New Haven, Conn.: Department of Anthropology, Yale University, 1972.

Conklin, Harold C., *Ethnographic Atlas of Ifugao: A Study of Environment, Culture, and Society in Northern Luzon.* New Haven, Conn.: Yale University Press, 1980.

Cook, O. F., "Milpa Agriculture: A Primitive Tropical System." In *Annual Report, 1919.* Washington, D.C.: Smithsonian Institution, 1919.

Cook, Sherburne F., *Soil Erosion and Population in Central Mexico.* Ibero-Americana 34. Berkeley and Los Angeles: University of California Press, 1949.

Coukis, Basil, principal coordinator, *Labor-Based Construction Programs: A Practical Guide for Planning and Management.* New Delhi: Oxford University Press, 1983.

Cressey, George B., "Qanats, Karez, and Foggaras." *The Geographical Review* 48(1958):27-44.

Criddle, Wayne D., and Cornelis Kalisvaart, "Subirrigation Systems." In *Irrigation of Agricultural Lands,* eds. Robert M. Hagan, Howard R. Haise, and Talcott W. Edminister. Agronomy Monographs 11. Madison: American Society of Agronomy, 1967.

Crossley, Peter, and John Kilgour, *Small Farm Mechanization for Developing Countries.* Chichester, England: Wiley, 1983.

Cruz López, Candido, "Estudio Agrológico Regional del Estado de Tlaxcala." *Ingeniería Hidráulica en México* 3, no. 1(1949):57-97, no. 3(1949):78-136, no. 4(1949):44-96.

Currier, Richard L., "The Hot-Cold Syndrome and Symbolic Balance in Mexican and Spanish-American Folk Medicine." *Journal of American Folklore* 66(1953):201–17.

Cushing, Frank Hamilton, *Zuñi Breadstuff.* New York: Museum of the American Indian, Indian Notes and Monographs 8(1920):165–66.

Dalrymple, Dana G., *Survey of Multiple Cropping in Less Developed Nations.* Foreign Economic Development Service 12. Washington, D.C.: U.S. Department of Agriculture, 1971.

Davies, J. W., *Mulching Effects on Plant Climates and Yield.* Technical Note 136 (WMO 388). Geneva: World Meteorological Organization, 1975.

de la Fuente, Julio, *Yalalag: Una Villa Zapoteca Serrana.* Serie Científica 1. Mexico City: Museo Nacional de Antropología, 1949.

de la Roca, Julio Cesar, *Bibliografía de un Pueblo: Síntesis Monográfica de Quezaltenango.* Guatemala City: Editorial "José de Pineda Ibarra," 1966.

de Lameiras, Brigitte B., *Terminología Agrohidráulica Prehispánica Nahua.* Colección Científica Historia 13. Mexico City: Instituto Nacional de Antropología e Historia, 1974.

Denevan, William M., "Aboriginal Drained-Field Cultivation in the Americas." *Science* 169(1970):647–54.

Denevan, William M., "Latin America." In *World Systems of Traditional Resource Management,* ed. Gary A. Klee. New York: V. H. Winston, 1980.

Denevan, William M., and B. L. Turner II, "Forms, Functions, and Associations of Raised Fields in the Old World Tropics." *Journal of Geography* 39(1974):24–33.

Diaz, Bernal, *The Conquest of Mexico,* trans. J. M. Cohen. Baltimore: Penquin Books, 1963.

Dillon, John L., Donald L. Plucknett, and Guy J. Vallaeys, *Farming Systems Research at the International Agricultural Research Centers.* Rome: Technical Advisory Committee to the Consultative Group on International Agricultural Research, Food and Agriculture Organization, 1978.

Dirección General de Conservación del Suelo y Agua (Mexico), *Resumen Gráfico de Trabajos de Control de la Erosión de los Suelos y de Conservación del Agua de las Lluvias.* Mexico City: Secretaría de Agricultura y Ganadaría, 1972.

Dirección General de Estadística (Guatemala), *II Censo Agropequario 1964.* 2 vols. Guatemala City: Ministerio de Economía, 1971.

Dirección General de Estadística (Guatemala), *VII Censo de Población y III de Habitación,* Ser. 1, vol. 1, *Población.* Guatemala City: Ministerio de Economía, 1973.

Dirección General de Estadística (Mexico), *IV Censo Agrícola, Ganadero, y Ejidal, 1960: Tlaxcala.* Mexico City: Secretaría de Industria y Comercio, 1965.

Dirección General de Estadística (Mexico), *IX Censo General de Población, 1970: Estado de Tlaxcala.* Mexico City: Secretaría de Industria y Comercio, 1971.

Donkin, R. A., *Agricultural Terracing in the Aboriginal New World.* Viking Fund Publications in Anthropology no. 56. Tucson: University of Arizona Press, Wenner-Gren Foundation for Anthropological Research, 1979.

Downing, Theodore L., "Irrigation and Moisture-Sensitive Periods: A Zapotec Case." In *Irrigation's Impact on Society,* eds. Theodore L. Downing and McGuire Gibson. Anthropological Papers of the University of Arizona 25. Tucson: University of Arizona Press, 1974.

Downing, Theodore L., and McGuire Gibson, eds., *Irrigation's Impact on Society.* Anthropological Papers of the University of Arizona 25. Tucson: University of Arizona Press, 1974.

Eckholm, Erik P., *Losing Ground.* New York: Norton, 1976.

Eckwall, E., "'Slash-and-Burn' Cultivation: A Contribution to Anthropological Terminology." *Man* 55(1955):135–36.

Enfield, Edward, *Indian Corn: Value, Culture, and Uses.* New York: Appleton, 1866.

English, Paul W., "The Origin and Spread of Qanats in the Old World." *Proceedings of the American Philosophical Society* 112(1968):170–81.

Erasmus, Charles J., "Monument Building: Some Field Experiments." *Southwestern Journal of Anthropology* 21(1965):277–301.

Escuela Facultativa de C. C. Económicas de Occidente, y Comité Interamericano de Desarrollo Agrícola, *Tenencia de la Tierra y Desarrollo Socio-Económico del Sector Agrícola en Guatemala.* 2d ed. Guatemala City: Editorial Universitaria, 1971.

Evans, L. J., "The Responses of Crops to Poor Soil Drainage." In *1971 Drainage,* ed. M. Hornung. Welsh Soils Discussion Group Report no. 12. Aberystwyth: Welsh Plant Breeding Station, 1971.

Evenari, M., L. Shanan, N. Tadmor, and Y. Aharoni, "Ancient Agriculture in the Negev." *Science* 133(1961):979–96.

Farvar, M. Taghi, and John P. Milton, *The Careless Technology: Ecology and International Development.* Garden City, N.Y.: Natural History Press, 1972.

Ferrari Bono, Bruno V., "Desarrollo de los Recursos Hidráulicos en los Valles Centrales de Oaxaca." Secretaría de Recursos Hidráulicos, Food and Agriculture Organization, Oaxaca City, 1968. Typescript.

Flannery, Kent V., "Archaeological Systems Theory and Early Mesoamerica." In *Anthropological Archaeology in the Americas.* Washington, D.C.: Anthropological Society of Washington, 1968.

Flannery, Kent V., Anne V. T. Kirkby, Michael J. Kirkby, and Aubrey W. Williams, Jr., "Farming Systems and Political Growth in Ancient Oaxaca." *Science* 158(1967):445–54.

Fluck, Richard C., and C. Direlle Baird, *Agricultural Energetics.* Westport, Conn.: AVI Publishing, 1980.

Folsom, Robert S., *Handbook of Greek Pottery.* Greenwich, Conn.: New York Graphic Society, 1967.

Food and Agriculture Organization, *Maize and Maize Diets.* Nutritional Studies 9. Rome: Food and Agriculture Organization, 1953.

Food and Agriculture Organization, *Organic Material as Fertilizers.* Soils Bulletin 27. Rome: Food and Agriculture Organization, 1975.

Food and Agriculture Organization, *Sandy Soils.* Soils Bulletin 25. Rome: Food and Agriculture Organization, 1975.

Food and Agriculture Organization, "Crop Area." In *Report on the 1970 World Census of Agriculture.* Statistics Series 10. Rome: Food and Agriculture Organization, 1977.

Food and Agriculture Organization, *1970 World Census of Agriculture: Analysis and International Comparison of the Results.* Rome: Food and Agriculture Organization, 1981.

Food and Agriculture Organization and United Nations Educational, Scientific, and Cultural Organization, *International Source-Book on Irrigation and Drainage of Arid Lands in Relation to Salinity and Alkalinity.* Rome: Food and Agriculture Organization and United Nations Educational, Scientific, and Cultural Organization, 1967.

Foster, George M., "Relationships Between Spanish and Spanish-American Folk Medicine." *Journal of American Folklore* 66(1953):201–17.

Foster, George M., *Culture and Conquest: America's Spanish Heritage.* Chicago: Quadrangle Books, 1960.

Fox, Roy L., John T. Phelan, and Wayne D. Criddle, "Design of Subirrigation Systems." *Agricultural Engineering* 37(1956):103–7.

Franklin, T. B., *A History of Agriculture.* London: G. Bell & Sons, 1948.

Freedman, Stephen M., "Human Labor as an Energy Source for Rice Production in the Developing World." *Agro-Ecosystems* 8(1982):125–36.

García, Enriqueta et al., *Precipitación y Probabilidad de Lluvia en el Estado de Oaxaca y su Evalución.* Mexico City: Comisión de Estudios del Territorio Nacional, and Instituto de Geografía, Universidad Nacional Autónoma de México, 1974.

García Casteñeda, C. Felipe, *Estudio Agrológico en la Zona Xochimilco-Chalco de la Cuenca del Valle de México.* Mexico City: Comisión Hidrológica de la Cuenca del Valle de México, Secretaría de Recursos Hidráulicos, 1962.

García Espinoza, Alfonso, "Estudio de los Suelos Tepetatosos y las Posibilidades de su Recuperación Agrícola." Thesis, Escuela Nacional de Agricultura, Chapingo, Mexico, 1960.

Gardner, W. R., "Soil Water Relations in Arid and Semiarid Conditions." In *Plant-Water Relationships in Arid and Semiarid Conditions*. Rome: United Nations Educational, Scientific, and Cultural Organization, 1960.

Geddes, Patrick, *Cities in Evolution*. New York: Howard Fertig, 1968 [1915].

Geertz, Clifford, *Agricultural Involution*. Berkeley and Los Angeles: University of California Press, 1963.

Geiger, Rudolf, *The Climate near the Ground*. Trans. of 4th German ed. Cambridge: Harvard University Press, 1966.

Gil Huerta, Gorgonio, "History of the Foundation of the Town of San Gabriel Chilacatla." In *The Prehistory of the Tehuacan Valley*, vol. 4, *Chronology and Irrigation*, ed. Frederick Johnson. Austin: University of Texas Press, 1972.

Glick, Thomas F., *Irrigation and Society in Medieval Valencia*. Cambridge: Harvard University Press, Belknap Press, 1970.

Gliessman, S. R., R. Garcia E., and M. Amador A., "The Ecological Basis for the Application of Traditional Agricultural Technology in the Management of Tropical Agro-Ecosystems." *Agro-Ecosystems* 7(1981):173–85.

Goe, Michael R., and Robert E. McDowell, *Animal Traction: Guidelines for Utilization*. International Agriculture Mimeograph 81. Ithaca, N.Y.: Cornell University, 1980.

Gollás, Manuel, "Mano de Obra Excedente y Eficiencia Económica en el Sector Tradicional de una Economía Dual: El Caso de Guatemala." *El Trimestre Económico* 40(1973):569–85.

Golomb, Berl, and Herbert M. Eder, "Landforms Made by Man." *Landscape* 4(1964):4–7.

Gómez-Pompa, Arturo, "An Old Answer to the Future." *Mazingira* (Pergamon Press) 5(1978):50–55.

Gouldner, Alvin W., and Richard A. Peterson, *Notes on Technology and the Moral Order*. Indianapolis: Bobbs-Merrill, 1962.

Grace, J., *Plant Response to Wind*. London: Academic Press, 1977.

Greenland, D. J., "Bringing the Green Revolution to the Shifting Cultivator." *Science* 190(1975):841–44.

Griffin, Ernst, and Howard W. Dennis, "A Mexican Corporate Campaign in Conservation." *Professional Geographer* 21(1969):358–59.

Grupo de Investigaciones Agrarias, *En el Centro de Documentación de Tecnología Campesina Tenemos*. Santiago: Proyecto de Tecnología Campesina y Organización, Academia de Humanismo Cristiano, 1983.

Guerra Borges, Alfredo, *Geografía Económica de Guatemala.* 2 vols. Guatemala City: Editorial Universitaria, 1969, 1973.

Hack, John T., *The Changing Physical Environment of the Hopi Indians of Arizona.* Papers of the Peabody Museum of American Archaeology and Ethnology, vol. 35, no. 1. Cambridge: Harvard University, 1942.

Hammond, Norman, "The Exploration of the Maya World." *American Scientist* 70(1982):482–495.

Haney, Emil B., Jr., "The Nature of Shifting Cultivation in Latin America." LTC 45. Madison: Land Tenure Center, University of Wisconsin, 1968. Mimeo.

Haring, Kirsten Johnson, *Preliminary Report on Field Work Carried Out in Hidalgo, Mexico, Between October 1973 and April 1974.* Worcester, Mass.: Graduate School of Geography, Clark University, [1974].

Harlan, Jack R., and Jean Pasquereau, "Decrue Agriculture in Mali." *Economic Botany* 23(1969):70–74.

Harper, J. L., "Agricultural Ecosystems." *Agro-Ecosystems* 1(1974):1-6.

Harris, Kenneth L., and Carl J. Lindblad, *Postharvest Grain Assessment Methods.* Washington, D.C.: American Association of Cereal Chemists, 1978.

Harrison, Peter D., and B. L. Turner II, eds., *Pre-Hispanic Maya Agriculture.* Albuquerque: University of New Mexico Press, 1978.

Haskins, Caryl P., *Of Ants and Men.* New York: Prentice-Hall, 1939.

Hatch, John K., *The Corn Farmers of Motupe: A Study of Traditional Farming Practices in Northern Coastal Peru.* Monograph 1. Madison: Land Tenure Center, University of Wisconsin, 1976.

Herold, Laurance C., *Trincheras and Physical Environment Along the Río Galiván, Chihuahua, Mexico.* Publications in Geography 65-1. Denver, Colo.: University of Denver, 1965.

Hewitt de Alcantara, Cynthia, *Modernizing Mexican Agriculture: Socioeconomic Implications of Technological Change 1940–1970.* Geneva: United Nations Research Institute for Social Development, 1976.

Hibbits, John E., personal communications. Servicios e Investigación OXFAM, Estación Experimental, Canton Choquí, Quezaltenango, Guatemala, 1974.

Holmes, R. M., and A. Nelson Dingle, "The Relationship Between the Macro- and Microclimate." *Agricultural Meteorology* 2(1965):127–33.

Hopfen, H. J., *Farm Implements for Arid and Tropical Regions.* Rev. ed. Agricultural Development Paper 91. Rome: Food and Agriculture Organization, 1969.

Hopfen, H. J., and E. Biesalski, *Small Farm Implements.* Agricultural

Development Paper 32. Rome: Food and Agriculture Organization, 1953.

Howard, William A., and Thomas M. Griffiths, *Trinchera Distribution in the Sierra Madre Occidental, Mexico.* Publications in Geography 66-1. Denver, Colo.: University of Denver, 1966.

Hunt, Robert C., and Eva Hunt, "Canal Irrigation and Local Social Organization." *Current Anthropology* 17(1976):389-411.

Hurst, G. W., and R. P. Rumney, *Protection of Plants Against Adverse Weather.* Technical Note 118 (WMO 281). Geneva: World Meteorological Organization, 1971.

ICRISAT (International Crops Research Institute for the Semi-Arid Tropics). *Proceedings of the International Workshop on Intercropping,* Hyderabad, India, Jan. 10–13, 1979.

Ingham, John M., "On Mexican Folk Medicine." *American Anthropologist* 72(1970):76–87.

Innis, Donald Q., "The Future of Traditional Agriculture." *Focus* 30, no. 3(Jan.–Feb. 1980):1–8.

Jacks, G. V., *Soil.* New York: Philosophical Library, 1954.

Jacks, G. V., R. Travernier, and D. H. Boalch, eds., *Multilingual Vocabulary of Soil Science.* Rome: Food and Agriculture Organization, 1960.

Jackson, I. J., "Traditional Forecasting of Tropical Rainy Seasons." *Agricultural Meteorology* 26(1982):167–78.

Jacobsen, Thorkild, and R. McC. Adams, "Salt and Silt in Ancient Mesopotamian Agriculture." *Science* 128(1958):1251–58.

Jaworski, Jerzy, "Evapotranspiration of Plants and Fluctuation of the Groundwater Table." In *Phreatophytes: A Bibliography,* vol. III-B. WRSIC 71-200. Washington, D.C.: Office of Water Resources Research, U.S. Department of the Interior, 1971.

Jeavons, John, *How to Grow More Vegetables (Than You Ever Thought Possible on Less Land Than You Can Imagine).* Palo Alto, Calif.: Ecology of the Midpeninsula, 1974.

Johnson, T. David, "Análisis de Actividades Necesarias para la Producción de 11 Especies Horticolas (Hortalizas)." Sector Público Agrícola—CETEFOR, Ministerio de Agricultura, Guatemala City, [1973]. Mimeo.

Johnson, Warren, *Muddling Toward Frugality.* San Francisco: Sierra Club Books, 1978.

Kalisvaart, C., *Subirrigation in the Zuiderzee Polders.* Wageningen, Netherlands: International Institute for Land Reclamation and Improvement, 1958.

Kanel, Don, "Size of Farm and Economic Development." *Indian Journal of Agricultural Economics* 22(1967):26–44.

Kedar, Yehuda, "Water and Soil from the Desert: Some Ancient Agricultural Achievements in the Central Negev." *Geographical Journal* 123(1957):179–87.

Keswani, C. L., and B. J. Ndunguru, *Proceedings of the Second Symposium on Intercropping in Semi-Arid Areas.* Morogoro, Tanzania: International Development Research Centre, 1982.

Khan, Amir U., "Appropriate Technologies: Do We Transfer, Adapt, or Develop?" In *Employment in Developing Nations,* ed. Edgar O. Edwards. New York: Columbia University Press, 1974.

Kimber, Clarissa, "Dooryard Gardens of Martinique." *Yearbook of the Association of Pacific Coast Geographers* 28(1966):97–118.

Kimber, Clarissa, "Spatial Patterning in the Dooryard Gardens of Puerto Rico." *The Geographical Review* 63(1973):6–26.

King, F. H., *Farmers of Forty Centuries.* Emmaus, Pa.: Rodale Press, 1973 [1911].

Kirkby, Anne V. T., *The Use of Land and Water Resources in the Past and Present Valley of Oaxaca, Mexico.* Memoirs of the Museum of Anthropology no. 5. Ann Arbor: University of Michigan Press, 1973.

Klee, Gary A., "Oceania." In *World Systems of Traditional Resource Management,* ed. Gary A. Klee. New York: V. H. Winston, 1980.

Klee, Gary A., ed., *World Systems of Traditional Resource Management.* New York: V. H. Winston, 1980.

Kostaninov, A. R., and L. R. Struzer, *Shelterbelts and Crop Yields (Lesnye polosy i urozhai).* Leningrad: Hydrometeorological Publishing House, 1965. (Jerusalem: Israel Program for Scientific Translations, 1969.)

Kostrowicki, Jerzy, "Agricultural Typology Concept and Method." *Agricultural Systems* 2(1977):33–45.

Latcham, Ricardo E., *La Agricultura Precolombiana en Chile y los Paises Vecinos.* Santiago, 1936.

Lathrap, Donald W., "Our Father the Cayman, Our Mother the Gourd: Spinden Revisited, or a Unitary Model for the Emergence of Agriculture in the New World." In *Origins of Agriculture,* ed. Charles A. Reed. The Hague: Mouton, 1977.

Leach, Gerald, *Energy and Food Production.* London: International Institute for Environment and Development, 1975.

Lees, Susan H., *Sociopolitical Aspects of Canal Irrigation in the Valley of Oaxaca.* Memoirs of the Museum of Anthropology no. 6. Ann Arbor: University of Michigan Press, 1973.

Lees, Susan H., "Oaxaca's Spiraling Race for Water." *Natural History* 84(1974):30–39.

Lele, Uma, "Rural Africa: Modernization, Equity, and Long-Term Development." *Science* 211(1981):547–53.

Léon Marquez, Borja, "Breve Estudio de las Condiciones Sociológicas e Higiénicas del Municipio de Sn. Gabriel Chilac, Tehuacán, Pue." Doctoral thesis, Facultad de Ciencias Médicas y Biológicas, Universidad Nacional de México, Mexico City, 1938.

Letey, J., L. H. Stolzy, and G. B. Blank, "Effect of Duration and Timing of Low Oxygen Content on Shoot and Root Growth." *Agronomy Journal* 54(1962):34-37.

Loomis, R. S., "Traditional Agriculture in America." *Annual Review of Ecology and Systematics* 15(1984):449-78.

Lorenzo, José L., "Aspectos Físicos del Valle de Oaxaca." *Revista Mexicana de Estudios Antropológicos* 16(1960):49-63.

Loucks, Orie L., "Emergence of Research on Agro-Ecosystems." *Annual Review of Ecological Systems* 8(1977):173-92.

McArthur, D., "Information Sources for Farmers from Outside Their Farms." *Agricultural Administration* 5(1978):275-80.

McBryde, Felix Webester, *Cultural and Historical Geography of Southwest Guatemala.* Institute of Social Anthropology Publication 4. Washington, D.C.: Smithsonian Institution, 1945.

Macedo Enciso, Miguel, *Manual del Magueyero.* Mexico City: Ediciones Agrícolas "Trucco," 1950.

MacNeish, Richard S., "Ancient Mesoamerican Civilization." *Science* 143(1964):531-37.

MacPherson, George, and Dudley Jackson, "Village Technology for Rural Development." *International Labour Review* 111(1975):97-118.

Madsen, Albert G., personal communication. Department of Agricultural and Natural Resource Economics, Colorado State University (formerly Asesor Técnico de Mercado Agrícola, El Salvador), 1975.

Matheny, Ray T., "Maya Lowland Hydraulic Systems." *Science* 193(1976):639-46.

Mather, Eugene, and John Fraser Hart, "The Geography of Manure." *Land Economics* 32(1956):25-38.

Mathewson, Kent, *Irrigation Horticulture in Highland Guatemala: The Tablón System of Panajachel.* Boulder, Colo.: Westview Press, 1984.

Merrill, William C., Lehman B. Fletcher, and Michael S. Hanrahan, "Vegetable Production and Marketing in Guatemala." U.S. Agency for International Development, Guatemalan Mission, 1971. Mimeo.

Miller, David H., *Water at the Surface of the Earth: An Introduction to Ecosystem Hydrodynamics.* International Geographical Series, vol. 21. New York: Academic Press, 1977.

Millikan, Max F., and David H. Hapgood, *No Easy Harvest.* Boston: Little, Brown, 1967.

Millon, René, Clara Hall, and May Díaz, "Conflict in the Modern Teotihuacán Irrigation System." *Comparative Studies in Society and History* 4(1961-62):494-524.

Molenaar, Aldert, *Water Lifting Devices for Irrigation.* Agricultural Development Paper 60. Rome: Food and Agriculture Organization, 1956.

Mondal, R. C., "Farming with a Pitcher." *World Crops* 26 (1974):94-97.

More, Rosemary J., "Water and Crops." In *Water, Earth, and Man: A Synthesis of Hydrology, Geomorphology, and Socio-Economic Geography,* ed. Richard J. Chorley. London: Methuen, 1969.

Mortensen, Ernest, and Ervin T. Bullard, *Handbook of Tropical and Sub-Tropical Horticulture.* Washington, D.C.: U.S. Agency for International Development, 1964.

Müller, J., "Labour-Intensive Methods in Low-Cost Road Construction: A Case Study." *International Labour Review* 101(1970):359-75.

Murray, D. B., "Crop-Weather Relations—Cacao." In *Agricultural Meteorology.* WMO 310. Geneva: World Meteorological Organization, 1972.

Nabhan, Gary Paul, "The Ecology of Floodwater Farming in Arid Southwestern North America." *Agro-Ecosystems* 5(1979):245-55.

Nash, Manning, *Primitive and Peasant Economic Systems.* San Francisco: Chandler, 1966.

National Academy of Sciences, *More Water for Arid Lands: Promising Technologies and Research Opportunities.* Washington, D.C.: National Academy of Sciences, 1974.

National Academy of Sciences, *Making Aquatic Weeds Useful: Some Perspectives for Developing Countries.* Washington, D.C.: National Academy of Sciences, 1976.

National Academy of Sciences, *Postharvest Food Losses in Developing Countries.* Washington, D.C.: National Academy of Sciences, 1978.

National Science Foundation, *Weather Modification, Tenth Annual Report, 1968.* Washington, D.C.: National Science Foundation, 1969.

Neghassi, Habte Mariam, "Effect of Smoothing on Furrow Irrigation." Master's thesis, Colorado State University, 1971.

Nietschmann, Bernard, *Between Land and Water: The Subsistence Ecology of the Miskito Indians, Eastern Nicaragua.* New York: Seminar Press, 1973.

Norman, David A., "Rationalizing Mixed Cropping Under Indigenous Conditions: The Example of Northern Nigeria." *Journal of Development Studies* 11(1974):3-21.

Norman, M.J.T., "Energy Inputs and Outputs of Subsistence Cropping Systems in the Tropics." *Agro-Ecosystems* 4(1978):355-66.

Norman, M.J.T., *Annual Cropping Systems in the Tropics.* Gainesville: University Presses of Florida, 1979.

Novoa B., Andrés R., and Joshua L. Posner, eds., *Agricultura de Ladera en América Tropical.* Turrialba, Costa Rica: Centro Agronómico Tropical de Investigación y Enseñanza, 1981.

Nye, P. H., and D. J. Greenland, *The Soil Under Shifting Cultivation.* Technical Communication 51. Farnham Royal, Bucks, England: Commonwealth Agricultural Bureaux, 1960.

Obregón Formoso, Leonardo, *Desarrollo de la Fruticultura por un Sistema Práctico y Económico de Riego en Zonas Aridas y Semi-Aridas.* Memorandum Técnico no. 318. Mexico City: Dirección General de Distritos de Riego, Secretaría de Recursos Hidráulicos, 1973.

Observatorio Nacional (Guatemala), *Datos Meteorológicos Mensuales Hasta 1959 Inclusiva.* Guatemala City: Observatorio Nacional, 1968.

Observatorio Nacional (Guatemala), *Datos Meteorológicos de las Cabeceras Departamentales.* Publicación Técnica 1. Guatemala City: Observatorio Nacional, 1972.

Oke, T. R., *Boundary Layer Climates.* London: Methuen, 1978.

Olquin Palacios, Carlos, and Francisco Oyarzabal Tamargo, "Consideraciones sobre Riego y Drenaje Agrícola en la Prehistoria de México." In *Agroecological Systems of Mexico: Contributions to Teaching, Research, and Agricultural Extension.* Chapingo: Escuela Nacional de Agricultura, 1977.

Orme, Bryony, "The Advantages of Agriculture." In *Hunters, Gatherers, and First Farmers Beyond Europe,* ed. J.V.S. Megaw. Leicester: Leicester University Press, 1977.

Ortiz, Sutti R., *Uncertainties in Peasant Farming.* New York: The Athlone Press, 1973.

Oshima, Harry T., "The Time to Change to Labour-Intensive Policies Is Now." *Ceres* 3(Nov.-Dec. 1970):31-34.

Owen, Wyn F., *Two Rural Sectors: Their Characteristics and Roles in the Development Process.* Occasional Paper 1. Bloomington: International Development Research Center, Indiana University, 1971.

Palerm, Angel, *Obras Hidráulicas Prehispánicas en el Sistema Lacustre del Valle de México.* Mexico City: Instituto Nacional de Antropología e Historia, 1973.

Palerm, Angel, and Eric Wolf, *Agricultura y Civilización en Mesoamérica.* Mexico City: Secretaría de Educación Público, 1972.

Parsons, James J., and William M. Denevan, "Pre-Columbian Ridged Fields." *Scientific American* 217(July 1967):92-100.

Parsons, Jeffrey R., "The Archaeological Significance of Mahamaes Cultivation on the Coast of Peru." *American Antiquity* 33(1968):80-85.

Parsons, Jeffrey R., and Norbert P. Psuty, "Sunken Fields and Prehistoric Subsistence on the Peruvian Coast." *American Antiquity* 40(1975):259-82.

Pasour, E. C., Jr., "Economic Efficiency: Touchstone or Mirage?" *Intercollegiate Review* 17(Fall-Winter 1981):33-44.

Pasour, E. C., Jr., and J. Bruce Bullock, "Energy and Agriculture: Some Economic Issues." In *Agriculture and Energy,* ed. William Lockeretz. New York: Academic Press, 1977.

Passmore, R., and J.V.G.A. Durnin, "Human Energy Expenditure." *Physiological Reviews* 35(1955):801-40.

Perrin, Richard K., Donald L. Winkleman, Edgardo R. Moscardi, and Jock A. Anderson, *From Agronomic Data to Farmer Recommendations: An Economics Training Manual.* Information Bulletin 27. Mexico City: Centro Internacional de Mejoramiento de Maíz y Trigo, 1976.

Phillips, P. G., "The Metabolic Cost of Common West African Agricultural Activities." *Journal of Tropical Medicine and Hygiene* 57(1954):12-20.

Pimental, David, et al., "Food Production and the Energy Crisis." *Science* 182(1973):443-49.

Pirie, N. W., "Using Plants Optimally." In *Food, Agriculture, and Environment,* eds. John Lenihan and William W. Fletcher. New York: Academic Press, 1976.

Price, Barbara J., "Pre-Hispanic Irrigation Agriculture in Nuclear America." *Latin American Research Review* 6, no. 3 (Fall 1971):3-60.

Primack, Martin L., "Farm Capital Formation as a Use of Farm Labor in the United States, 1850-1910." *Journal of Economic History* 26(1966):348-62.

Prohaska, F., "Weather Modification in the Tropics for Agricultural Purposes." In *Agricultural Meteorology.* WMO 310. Geneva: World Meteorological Organization, 1972.

Radke, J. K., and R. T. Hagstrom, "Wind Turbulence in a Soybean Field Sheltered by Four Types of Wind Barriers." *Agronomy Journal* 66(1974):273-78.

Ramírez P., Felix, and R. J. Laird, *Densidad Optima de Plantas de Maíz para los Valles de México y Toluca.* Folleta Técnica 42. Mexico City: Oficina de Estudios Especiales, Secretaría de Agricultura y Ganadaría, 1960.

Redfield, Robert, *The Little Community.* Chicago: University of Chicago Press, 1956.

Reina, Ruben E., *The Law of the Saints: A Pokoman Pueblo and Its Community Culture.* Indianapolis: Bobbs-Merrill, 1966.

Rojas, Rabiela T., *Aspectos Tecnológicos de las Obras Hidráulicas Prehispánicas y Coloniales en el Valle de México.* Mexico City: Instituto Nacional de Antropología e Historia, 1974.

Rosales Padilla, José Luis, *Evaluación de Sequías en la República de Guatemala con Base en Datos de Precipitación Mensual.* Guatemala City: Instituto Geográfico Nacional, 1967.

Rose, G. J., *Crop Protection.* New York: Philosophical Society, 1955.

Rosenberg, Norman J., *Microclimate: The Biological Environment.* New York: Wiley, 1974.

Ruthenberg, Hans, *Farming Systems in the Tropics,* 2d ed. Oxford: Clarendon Press, 1976.

Ruttan, Vernon W., *Agricultural Research Policy.* Minneapolis: University of Minnesota Press, 1982.

Sahagún, Bernardino de, *General History of the Things of New Spain (Florentine Codex),* trans. Arthur O. Anderson and Charles E. Dibble. Monographs of the School of American Research and the Museum of New Mexico no. 14. Santa Fe: School of American Research, pt. 4, 1952; pt. 12, 1963; pt. 13, 1963.

Saks, Margaret E., and C. Ronald Carroll, "Ant Foraging Activity in Tropical Agro-Ecosystems." *Agro-Ecosystems* 6(1980):177–88.

Salter, R. M., and C. J. Schollenberger, *Farm Manure.* Bulletin 605. Ohio Agricultural Experiment Station, 1939.

Sanchez, James, "Qanats as a Viable Contemporary Irrigation Technology." In *Tecnología Apropriada para Paises Sub-Desarrollados.* San Salvador City: Universidad Centroamericana José Simeon Canas, 1979.

Sanders, William T., "Tierra y Agua (Soil and Water): A Study of the Ecological Factors in the Development of Meso-American Civilization." Ph.D. diss., Harvard University, 1957.

Sauer, Carl O., and Donald Brand, *Prehistoric Settlements of Sonora with Reference to Cerros de Trincheras.* Publications in Geography 5. Berkeley: University of California, 1931.

Schilling, Elisabeth, *Die "Schwimmenden Gärten" von Xochimilco.* Schriften des Geographischen Institutes, vol. 9, no. 3. Kiel: Universität Kiel, 1938.

Schmid, Lester, *El Papel de la Mano de Obra Migratoria en el Desarrollo Económico de Guatemala.* Guatemala City: Instituto de Investigaciones Económicas y Sociales, Universidad de San Carlos, 1973.

Schultz, Theodore W., *Transforming Traditional Agriculture.* New Haven, Conn.: Yale University Press, 1964.

Schuman, Gerald E., and Robert E. Burwell, untitled notice. *Water Newsletter* (Water Information Center) 15(Dec. 12, 1973).

Schwab, Glen O., Richard K. Frevert, Kenneth K. Barnes, and Talcott W. Edminister, *Elementary Soil and Water Engineering,* 2d ed. New York: Wiley, 1971.

Secretaría de Obras Públicos (Mexico), "Obras a Mano." *Organo Informativo SOP* 83(Feb. 1973).

Secretaría de Recursos Hidráulicos (Mexico), "Obras en el Estado de Tlaxcala; Distrito de Riego por Bombeo de Zacatelco." Tlaxcala, Mexico, n.d. Typescript.

Seele, Enno, "Galerías Filtrantes en el Area de Acatzingo-Tepeaca, Estado de Puebla." *Instituto Nacional de Antropología e Historia Boletin* (Mexico City) 35(1969):3-8.

Seele, Enno, "Galerías en el Estado de Puebla." *Comunicaciones Proyecto Puebla-Tlaxcala* (Fundación Alemana para la Investigación Científica) 7(1973):141-44.

Shaikh, Rashid, "Commentary: Reflections on Appropriate Science in India." *Science for the People* 13(1981):26-27.

Shattuck, George Cheever, *The Peninsula of Yucatan: Medical, Biological, Meteorological, and Sociological Studies.* Publication 431. Washington, D.C.: Carnegie Institution, 1933.

Simmons, Charles S., José Manuel Tárano T., and José Humberto Pinto Z., *Clasificación de Reconocimiento de los Suelos de la República de Guatemala.* La Aurora, Guatemala: Instituto Agropequario Nacional, Ministerio de Agricultura, 1959.

Smith, Allan E., and D. M. Secoy, "Salt as a Pesticide, Manure, and Seed Steep." *Agricultural History* 50(1976):506-16.

Smith, C. Earle, Jr., "Agriculture, Tehuacán Valley." *Fieldiana: Botany* 31, no. 3(1965):52-100.

Soil Conservation Service, *A Manual on Conservation of Soil and Water.* Agriculture Handbook no. 61. Washington, D.C.: U.S. Department of Agriculture, 1954.

Spencer, J. E., and G. A. Hale, "The Origin, Nature, and Distribution of Agricultural Terracing." *Pacific Viewpoint* 2(1961):1-40.

Spores, Ronald, "Settlement, Farming Technology, and Environment in the Nochixtlan Valley." *Science* 166(1969):557-69.

Stanislawski, Dan, *Landscapes of Bacchus.* Austin: University of Texas Press, 1970.

Steensberg, Axel, "Comments" in Robert C. Hunt and Eva Hunt, "Canal Irrigation and Local Social Organization." *Current Anthropology* 17(1976):389-411.

Stephens, John L., *Incidents of Travel in Yucatan.* New York: Harper, 1848.

Stevens, Rayfred L., "The Soils of Middle America and Their Relation to the Indian Peoples and Cultures." In *Handbook of Middle American Indians,* vol. 1, *Environment and Early Cultures,* ed. Robert C. West. Austin: University of Texas Press, 1964.

Steward, Julian H., "Irrigation Without Agriculture." *Papers of the Michigan Academy of Science, Arts, and Letters* 12(1930):149-56.

Stewart, Omer C., "Fire as the First Great Force Employed by Man." In *Man's Role in Changing the Face of the Earth,* ed. William L. Thomas, Jr. Chicago: University of Chicago Press, 1956.

Stewart, W.D.P., "Nitrogen-Fixing Plants." *Science* 158(1967):1426-32.

Stigter, C. J., *Microclimate Management and Manipulation in Traditional Farming*. Report of a Working Group, 8th sess. Geneva: Commission for Agricultural Meteorology, World Meteorological Organization, 1983.

Stigter, C. J., "Shading: A Traditional Method of Microclimate Manipulation." *Netherlands Journal of Agricultural Science* 32(1984):81-86.

Sullivan, Richard H., Morris M. Cohn, and Samuel S. Baxter, *Survey of Facilities Using Land Application of Wastewater*. EPA-430/9-73-006. Washington, D.C.: Office of Water Program Operations, Environmental Protection Agency, 1973.

Sutton, A. L., J. V. Mannering, D. H. Bache, J. F. Marten, and D. D. Jones, *Utilization of Animal Waste as Fertilizer*. Cooperative Extension Service ID-101. West Lafayette, Ind.: Purdue University, 1975.

Tamayo, Jorge L., *Geografía General de México*, 2d ed., 4 vols. Mexico City: Instituto Mexicano de Investigaciones Económica, 1962.

Tax, Sol, and Robert Hinshaw, "The Maya of the Midwestern Highlands." In *Handbook of Middle American Indians*, vol. 7, *Ethnology, Part One*, ed. Evon Z. Vogt. Austin: University of Texas Press, 1969.

Thorne, D. W., and H. B. Peterson, *Irrigated Soils: Their Fertility and Management*. Philadelphia: Blakiston, 1949.

Thornthwaite, C. W., and J. R. Mather, *Instructions and Tables for Computing Potential Evapotranspiration and the Water Balance*. Publications in Climatology, vol. 10, no. 3. Centerton, N.J.: Laboratory of Climatology, Drexel Institute of Technology, 1957.

Tindall, H. D., *Commercial Vegetable Growing*. London and Ibadan, Nigeria: Oxford University Press, 1968.

Tolman, C. L., *Ground Water*. New York: McGraw-Hill, 1937.

Trafford, D. B., "Agricultural Land Drainage." In *1971 Drainage*, ed. M. Hornung. Welsh Soils Discussion Group Report no. 12. Aberystwyth: Welsh Plant Breeding Station, 1971.

Tresemer, David, *Transplants in Soil Blocks*, rev. ed. Brattleboro, Vt.: By Hand & Foot, 1985.

Troll, Carl, "Qanat-Bewässerung in der Alten und Neuen Welt." *Mitteilungen der Osterreichischen Geographischen Gesellschaft* 105(1954):313-30.

Turner, B. L., II, "Prehistoric Intensive Agriculture in the Mayan Lowlands." *Science* 185(1974):118-24.

Turner, B. L., II, Robert Q. Hanham, and Anthony V. Portararo, "Population Pressure and Agricultural Intensity." *Annals of the Association of American Geographers* 67(1977):384-96.

Ucko, Peter J., and G. W. Dimbleby, eds., *The Domestication and Exploitation of Plants and Animals.* Chicago: Aldine, 1969.

van Eimern, J., R. Karschon, L. A. Razumova, and G. W. Robertson, *Windbreaks and Shelterbelts.* Technical Note 59 (WMO 147). Geneva: World Meteorological Organization, 1964.

Van Slyke, Lucius L., *Fertilizers and Crops.* New York: Orange Judd, 1912.

Ventskevich, G. Z., *Agrometeorology.* Leningrad: Hydrometeorological Publishing House, 1958. (Jerusalem: Israel Program for Scientific Translations, 1961.)

Villegas, Bernardo M., "For Appropriate Technology That Uses the Resources of the Third World by Substituting Labour for Capital." *Ceres* 7(May-June 1974):44-47.

Vogt, Evon Z., *Zinacantan: A Maya Community in the Highlands of Chiapas.* Cambridge: Harvard University Press, 1969.

Volunteers in Technical Assistance, "The Amazing Windmills of Miramar." *VITA News* 2(July 1982):8-10.

Waddell, Eric, "How the Enga Cope with Frost: Responses to Climatic Perturbations in the Central Highlands of New Guinea." *Human Ecology* 3(1975):249-73.

Wagner, Philip L., *Nicoya: A Cultural Geography.* Publications in Geography, vol. 12, no. 3. Berkeley: University of California, 1958.

Ward, G. M., T. M. Sutherland, and J. M. Sutherland, "Animals as an Energy Source in Third World Agriculture." *Science* 208(1980):570-74.

Watters, R. F., *Shifting Cultivation in Latin America.* Rome: Food and Agriculture Organization, 1971.

Weathers, Kenneth, "La Agricultura de los Tzotzil de Nobenchauc, Chiapas, México." *América Indígena* 6(1946):315-19.

Weatherwax, Paul, *Indian Corn in Old America.* New York: Macmillan, 1954.

Weaver, John E., *Root Development of Field Crops.* New York: McGraw-Hill, 1926.

Weber, Neal A., *Gardening Ants: The Attines.* Philadelphia: American Philosophical Society, 1972.

Welchert, W. T., and Barry N. Freeman, "'Horizontal' Wells." *Journal of Range Management* 26(1973):253-55.

Werner, Gerd, Hans Aeppli, Günter Miehlich, and Ernst Schönhals, "Los Suelos de la Cuenca Alta de Puebla-Tlaxcala y sus Alrededores." *Comunicaciones* (Fundación Alemana para la Investigación Científica) 6(Supp.)(1978):1-95.

Wesseling, J., et al., "Land Drainage in Relation to Soils and Crops." In *Drainage in Agricultural Lands,* ed. James N. Luthin. Madison, Wis.: American Society of Agronomy, 1957.

West, Michael, "Early Watertable Farming on the North Coast of Peru." *American Antiquity* 44(1979):138-44.

West, Robert C., *Cultural Geography of the Modern Tarascan Area.* Institute of Social Anthropology 7. Washington, D.C.: Smithsonian Institution, 1948.

West, Robert C., "Population Densities and Agricultural Practices in Pre-Columbian Mexico, with Emphasis on Semi-Terracing." *Verhandlungen des XXXVIII Internationalen Amerikanistenkongresses* 2(Aug. 12-18, 1968):361-69.

West, Robert C., and Pedro Armillas, "Las Chinampas de México: Poesia y Realidad de los 'Jardines Flotantes.'" *Cuadernos Americanos* 50(1950):165-82.

West, Robert C., and John R. Augelli, *Middle America: Its Lands and Peoples,* 2d ed. Englewood Cliffs, N.J.: Prentice-Hall, 1976.

Wharton, Clifton R., Jr., "Research Priorities on Subsistence Agriculture." In *Subsistence Agriculture and Economic Development,* ed. Clifton R. Wharton, Jr. Chicago: Aldine, 1969.

Wheatly, Paul, "Agricultural Terracing." *Pacific Viewpoint* 5(1965):123-44.

Whiteford, Scott, and Luis Emilio Henao, "Commercial Agriculture, Irrigation Control, and Selective Labor Migration: The Case of the Tehuacán Valley." Paper presented at the American Anthropological Association meetings, Houston, Texas, 1978. Mimeo.

Whiteford, Scott, and Luis Emilio Henao, "Irrigación Descentralizado, Desarrollo y Cambio Social." *America Indígena* 40(1980):57-72.

Whyte, R. O., G. Nilsson-Leissner, and H. C. Trumble, *Legumes in Agriculture.* Agricultural Studies 21. Rome: Food and Agriculture Organization, 1953.

Wilken, Gene C., "Drained-Field Agriculture: An Intensive Farming System in Tlaxcala, Mexico." *The Geographical Review* 59(1969):215-41.

Wilken, Gene C., "The Ecology of Gathering in a Mexican Farming Region." *Economic Botany* 24(1970):286-95.

Wilken, Gene C., "Food-Producing Systems Available to the Ancient Maya." *American Antiquity* 36(1971):432-48.

Wilken, Gene C., "Microclimate Management by Traditional Farmers." *The Geographical Review* 62(1972):544-60.

Wilken, Gene C., "Integrating Forest and Small-Scale Farm Systems in Middle America." *Agro-Ecosystems* 3(1977):291-302.

Wilken, Gene C., "Manual Irrigation in Middle America." *Agricultural Water Management* 1(1977):155-65.

Wilken, Gene C., "Traditional Slope Management: An Analytical Approach." In *Hill Lands: Proceedings of an International Symposium.* Morgantown: West Virginia University Books, 1979.

Wilken, Gene C., "The Fallacy of Labour/Land Comparisons of Agricultural Productivity." In *Perspectives in Agricultural Geography,* vol. 4, ed. Noor Mohammad. New Delhi: Concept Publishing, 1980.

Wilken, Gene C., *Agroclimatic Hazard Perception, Prediction, and Risk-Avoidance Strategies in Lesotho.* Natural Hazard Research Working Paper no. 44. Boulder: University of Colorado, 1982.

Wilken, Gene C., "A Note on Buoyancy and Other Dubious Characteristics of the 'Floating' Chinampas of Mexico." *British Archaeological Reports (B.A.R.),* International Ser. 232(1985):31-48.

Wilkinson, Albert E., *Muck Crops.* New York: Orange Judd, 1916.

Williams, Barbara J., "Tepetate in the Valley of Mexico." *Annals of the Association of American Geographers* 62(1972):618-26.

Williams, Barbara J., "Aztec Soil Science." *Boletín, Instituto de Geografía* (Universidad Nacional Autónomo de México) 6(1976):115-20.

Williams, Barbara J., "Middle American Folk Soil Taxonomy." *Annals of the Association of American Geographers* 71(1981):337-58.

Williams, Barbara J., "Pictorial Representation of Soils in the Valley of Mexico: Evidence from the Codex Vergara." *Geoscience and Man* 21(1981):51-62.

Winterhalder, Bruce, and Eric Alden Smith, *Hunter-Gatherer Foraging Strategies.* Chicago: University of Chicago Press, 1981.

Wischmeier, W. H., and J. V. Mannering, "Relation of Soil Properties to Its Erodibility." *Proceedings of the Soil Science Society of America* 33(1969):131-37.

Wittfogel, Karl A., "The Hydraulic Approach to Pre-Spanish Mesoamerica." In *The Prehistory of the Tehuacan Valley,* vol 4, *Chronology and Irrigation,* ed. Frederick Johnson. Austin: University of Texas Press, 1972.

Woodbury, Richard B., "Prehistoric Water Management Systems in the Tehuacán Valley, Mexico." In *XXXVI Congreso Internacional de Americanistas,* vol. 1. Mexico City: Editorial Comaval, 1965.

World Bank, *World Bank Study of the Substitution of Labor and Equipment in Civil Construction.* Technical Memoranda 1-25. Washington, D.C.: International Bank for Reconstruction and Development, 1975-76.

No. 3, "Comparison of Different Modes of Haulage in Earthworks," June 1975.

No. 11, "A Literature Review of the Ergonomics of Labor Intensive Civil Construction," Aug. 1975.

No. 12, "Haulage by Headbaskets, Shoulder Yokes, and Other Manual Load Carrying Methods," Oct. 1975.

No. 18, "Spreading Activities in Civil Construction," Dec. 1975.

No. 19, "Excavation," Feb. 1976.

No. 20, "Loading and Unloading Activities," Feb. 1976.

No. 21, "A Literature Review of the Work Output of Animals with Particular Reference to Their Use in Civil Construction," Feb. 1976.

World Bank, *World Development Report 1980*. New York: Oxford University Press, 1980.

Wright, A.C.S., "Some Terrace Systems of the Western Hemisphere and Pacific Islands." *Pacific Viewpoint* 3(1962):97-101.

Zaretske, Kenneth, "The Idea of Tradition." *The Intercollegiate Review* (Spring/Summer 1982):85-96

Index

Designer: Trinity Publishers Services
Compositor: WESType Publishing Services, Inc.
Text: 10/12 Baskerville
Display: Baskerville